Bhadresh R Sudani

Chroniques chimiques du Moringa Oleifera : dévoiler l'alchimie de la nature

Bhadresh R Sudani

Chroniques chimiques du Moringa Oleifera : dévoiler l'alchimie de la nature

Le mystère miraculeux du Moringa

ScienciaScripts

Imprint

Any brand names and product names mentioned in this book are subject to trademark, brand or patent protection and are trademarks or registered trademarks of their respective holders. The use of brand names, product names, common names, trade names, product descriptions etc. even without a particular marking in this work is in no way to be construed to mean that such names may be regarded as unrestricted in respect of trademark and brand protection legislation and could thus be used by anyone.

Cover image: www.ingimage.com

This book is a translation from the original published under ISBN 978-620-7-48725-7.

Publisher:
Sciencia Scripts
is a trademark of
Dodo Books Indian Ocean Ltd. and OmniScriptum S.R.L publishing group

120 High Road, East Finchley, London, N2 9ED, United Kingdom
Str. Armeneasca 28/1, office 1, Chisinau MD-2012, Republic of Moldova, Europe
Printed at: see last page
ISBN: 978-620-8-13644-4

PROLOGO

The liver is an organ in which the probability of having a secondary tumor is 18 to 40 times greater than a primary tumor. The pattern of appearance of metastases in the liver depends on the location of the primary tumor, its histologic type, as well as the sex and age of the patient. Approximately half of the metastases that settle in the liver come from the gastrointestinal tract, and within this group, cancer of the colon and rectum is the tumor that most frequently produces metastases in the liver.

The number of cancer cases in Spain is increasing every year. Of the total, 3,000 new cases diagnosed correspond to liver cancer. More than 50% of these patients are candidates for treatment with minimally invasive thermoablation techniques, a methodology which, according to the latest studies, improves the recurrence rates of the disease, as well as the patients' expectations and quality of life. However, only half of the patients who could be candidates have access to this type of therapy. The precision of this technology also stands out in its indication for liver metastasis, a pathology that appears, for example, in 50% of patients with colorectal cancer, one of the most common cancers in Spain.

The increasing incidence of liver tumors has promoted the development of different minimally invasive ablative therapeutic options that are gaining special relevance in clinical practice as a complement or alternative to surgery. In this scenario it is important to know the indications of each of them in order to personalize the treatment for each patient.

INDEX

Chapter 1: General information and types of ablation

INTRODUCTION

The number of cancer cases in Spain is increasing every year. Of the total, more tan 3,000 new diagnoses correspond to liver cancer. More than 50% of these patients are candidates for treatment with minimally invasive thermoablation techniques, a methodology which, according to the latest studies, improves the recurrence rates of the disease, as well as patients' expectations and quality of life.

Thermal ablation is a minimally invasive, image-guided treatment that uses extreme heat or cold to destroy cancerous tumor cells.

Local tumor ablation techniques, also known as image-guided tumor ablation techniques, are a group of procedures in which the "therapeutic stimulus" is selectively administered in the tumor area while preserving the rest of the non-tumorous parenchyma unharmed. This usually implies causing by different physicochemical methods an area of tumor necrosis while preserving the surrounding normal tissue (1).

In the past two decades, ablative therapies with chemical compounds or thermal energy have emerged as an effective treatment for the curative management of small primary or metastatic liver tumors. (2)

Complete tumor ablation and cure of the patient may be a possible goal in selected cases. In other cases an increase in survival or symptomatic improvement can be achieved. In hepatic pathology the route of administration of this stimulus is frequently percutaneous, but occasionally the access is surgical, through laparotomy or laparoscopy with/without liver resection, and can offer considerable advantages. Two main groups of techniques can be distinguished:

• Techniques that apply chemical treatment: Percutaneous ethanol injection (PEI). This consists of the injection of pure ethanol inside the lesion to be treated, producing cellular dehydration, denaturation of proteins and finally cellular apoptosis. It has been described that this technique is not useful for metastasis, although it is considered as effective as the other techniques for the treatment of hepatocarcinoma smaller than 2 cm.

- Techniques that apply heat treatment: There are several methods of ablative therapies (3):

- Radiofrequency (RF): use of local heat. The aim is to induce coagulative necrosis by means of electromagnetic alternating current in a frequency range of approximately 375-500 kHz to achieve a temperature of 90-120ºC.

- Microwaves (microwave= MW) It is also about inducing a necrosiscoagulation through a microwave emitter that generates electromagnetic energy with frequencies equal to or higher than 900 kHz.

- Cryotherapy: laser or the use of local cold by means of cryosurgery (CC) It consists of the formation of intracellular ice crystals that induce tissue necrosis. Needles are used which, by means of circulating gas expansion systems, reach temperatures of up to -40ºC and thus irreversibly destroy the tissue.

- Electroporation It is a technique based on the use of electrical pulses that cause defects in the cell membrane at the nanometer level, called nanopores or conductive pores, generating a permeation of the membrane of the target cells. This permeation can be temporary (reversible electroporation) but can also be permanent above a certain electrical level, causing cell death by disruption of cell homeostasis (irreversible electroporation).

- Application of ultrasound, (usually in the form of high-intensity focused ultrasound (HIFU))

- Nanothermis/Oncothermia

- Laser Interstitial Thermal Therapy (LITT)

All of them are considered minimally invasive techniques and have at least three advantages over conventional resective surgical procedures:

1. The possibility of treating patients without surgical indication.

2. Reduced morbidity and mortality and better patient tolerance.

3. Reduced economic cost per procedure.

Thermal ablation, as its name suggests, is based on the generation of heat within the target tissue to achieve tissue destruction. The aim of thermal ablation is to heat malignant tissues to temperatures that can induce immediate coagulative necrosis (usually above 60°C (4).

The liver can be particularly difficult to treat effectively, as this organ has high tissue perfusion and large blood vessels that can act as "heat sinks" near the ablation site.

Such heat sinks can decrease the effect of treatment on malignant cells to sub-lethal temperature levels, increasing the likelihood of local tumor progression by limiting the size and efficacy of the ablation site.

A complete treatment should encompass the tumor plus a 5 to 10 mm margin (analogous to a surgical margin), but respecting healthy parenchyma and vulnerable structures. (5,6) Optimal treatment requires:

- Proper placement of the device
- Sufficient power supply
- Verification of ablation margins

The different ablation methods are summarized in Table 1.

Table 1: Ablation methods and mechanism of action

Radiofrequency (RFA) Heat	High-frequency electric current (350-500 KHz range) passes through an electrode, creating frictional heat that destroys tissues and cells. Direct heating by FRA occurs within a few millimeters of the needle, but a final ablation zone is created when thermal conduction pushes heat to more peripheral areas around the electrode.
Microwave (MWA) Heat	A form of thermal ablation that uses electromagnetic waves at frequencies in the microwave energy spectrum (915 MHz or 2.45 GHz in ablative techniques) to produce tissue heating effects. In tissue, heating occurs because the electromagnetic (EM) field forces water molecules to oscillate. The bound water molecules tend to oscillate out of phase, so some of the EM energy is absorbed and converted into heat.
Laser (LITT) Heat	Laser-induced thermal therapy (LITT) is a percutaneous tumor ablation technique that uses high-power lasers placed interstitially in the tumor to deliver therapy. Modern systems use small, compact, high-powered diode laser systems with actively cooled applicators to help prevent tissue charring during the procedure.
High Intensity Focused Ultrasound (HIFU) Heat	It is a technology that uses ultrasound waves for tissue ablation. HIFU therapy can transport energy in the form of ultrasound waves through different tissues to a given object. It produces an increase in temperature (thermal effect) and other biological interactions, the most significant being acoustic cavitation, in an absolutely non-invasive manner. It is the only procedure that does not require the insertion of a needle or antenna inside the patient's body.
Cryoablation Cold	It is a treatment to kill cancer cells with extreme cold. During cryoablation, a thin, rod-shaped needle (cryoprobe) is inserted directly into the tumor. A gas (liquid nitrogen or argon gas) is pumped into the cryoprobe to create intense cold to freeze and destroy the diseased tissue, and then the tissue is allowed to thaw. The freezing and thawing process can be repeated several times during the same treatment session.
Irreversible electroporation (IRE) Non-thermal	It is a tissue ablation technique that uses ultra-short but strong electric fields to create permanent and lethal nanopores in the cell membrane to disrupt cell homeostasis. IRE applies several series of electrical pulses (90 pulses) from 1500 to 3000 V. Unlike thermal ablation, IRE affects only the cell membrane while preserving the extracellular scaffold, so lumen structures such as blood vessels, bile ducts and intestine remain patent and can regenerate.

Advances in navigation systems, augmented reality, 3D reconstruction, control of results and new ablation systems will make percutaneous ablation (80%) or laparoscopic ablation (20%) the *gold standard* for tumors of 3cm or less in the liver in the near future; the latter in peripheral tumors or in combination with surgical treatments. Tumor ablation techniques in the treatment of hepatic tumors have gained ground in palliative and especially therapeutic management (Table 2).

Table 2: indications for each liver ablation method

IN LIVER	RADIO FREQUENCY	MICROWAVE	IRE	CRYOTHERAPY
Tumors<2cm	+++	++	++	++
Tumors>2cm	++	+++	+	++
Tumors near peripheral vascular structures	+	+++	++	++
Tumors near central vascular structures or biliary tract	++	+	+++	+

There are two phenomena to take into account that can eliminate or reduce the thermal effect of ablation devices:

- Heat sink effect: Refers to the cooling effect of the vessels adjacent to the ablation area with a diameter >1 mm. This effect conditions a protection of the blood vessels, although it would also condition a possible decrease in the theoretical ablation area.
- Perfusion-mediated tissue cooling: Refers to the cooling effect of capillary

microperfusion effects on tissue. There are multiple techniques to reduce this effect based on pharmacotherapy or invasive techniques such as temporary balloon vascular occlusion, definitive vascular occlusion (embolization) and surgical vascular occlusion.

Percutaneous ablation techniques are minimally invasive procedures, although they are not risk-free. Therefore, the potential complications and associated mortality must be known in order to assess the risks and benefits on an individual basis. There is great heterogeneity in the studies of these techniques, making it difficult to establish quality standards, ablation complications, success rates and disease recurrence after ablation. A major complication is defined as all those symptoms that appear after ablation and persist for more than one week, that delay hospital discharge, that are associated with important comorbidity or that compromise the patient's life.

SIR (Society of Interventional Radiology) classification to categorize the severity of the lesions (Table 3):

Table 3: SIR classification of complication severity

Categoría	Definición
I	Sin tratamiento, consecuencias o secuelas adversas.
II	Requiere un incremento no planeado del nivel de cuidado, o mínima consecuencia o secuela.
III	Requiere un incremento no planeado del nivel de cuidado a un grado intermedio o una secuela intermedia, con hospitalización en el hospital para observación.
IV	Requiere un incremento no planeado del nivel de cuidado a un grado mayor, con secuela grave y hospitalización prolongada (>48 horas).
V	Muerte directa o indirectamente relacionada con el procedimiento.

The rate of death and serious complications after thermal ablation are low:

- Mortality: 0 -1.4%.
- Major complications: 2.2 -5.7% (7)

TABLE 4. Complications after ablation of hepatic lesions

Causas de muerte	Perforación intestinal	
	Trombosis portal	
	Fallo hepático	
	Shock séptico	
	Hemorragia hepática masiva	
Complicaciones mayores	Hemorragia	
	Fallo hepático	
	Complicaciones intestinales	
	Complicaciones biliares	Estenosis biliar
		Bilioma
		Colecistitis
		Fistulas broncobiliares
	Infecciones	Abscesos
		Peritonitis
	Trombosis vascular e infarto hepático	
	Complicaciones pleurales	Pneumotorax
		Hemotorax
		Derrame pleural masivo
Complicaciones menores	Dolor	
	Fiebre	
	Derrame pleural asintomático.	

Vascular complications

Hemorrhage: Hemorrhage is one of the most frequent major complications related to hepatic thermal ablation. Overall the risk of bleeding is low (<2%), although it may vary depending on the degree of cirrhosis of the liver parenchyma and the location of the tumor (e.g., if the tumor is close to a large vessel). Bleeding is usually intraperitoneal, but may also be subcapsular, intralesional, intraparenchymal, or pleural. There may be 3 hemorrhagic conditions: active arterial bleeding, pseudoaneurysm, and sheet venous bleeding. Most bleeds are usually limited and can be treated conservatively. Active arterial bleeding and pseudoaneurysm will require endovascular treatment.

To prevent bleeding should be taken into account:

- Correction of coagulopathies.
- Transfer the hepatic capsule as few times as possible.
- Avoid large vessels in the needle path.
- Cauterization of the needle track has been shown to reduce the risk of bleeding.

Vascular fistulas: There may be pathologic scarring conditioning an arterio-portal or arterio-venous communication (with suprahepatic vein), arterio-biliary or intestinal communication. Their treatment usually requires selective embolization and less frequently surgical correction.

Portal and suprahepatic thrombosis: The incidence of portal and suprahepatic vein thrombosis is between 1.7% and 1.4% respectively. Portal thrombosis usually occurs rapidly after the procedure. Small caliber vessels (<3 mm) tend to be more susceptible to thrombosis from thermal injury as they do not have heat-sink effect (only large vessels). Thermal injury can cause thrombosis in large vessels if there is decreased flow (e.g., in cirrhotic patients). Treatment of venous thrombosis is assessed on an individual basis. It usually does not require treatment, although systemic anticoagulation or local thrombolysis may be required if liver function is impaired.

Hepatic infarction: This is an infrequent complication (1.8%) due to the double hepatic blood supply and the important vascular collaterality. When hepatic infarction is established, treatment is usually conservative.

1. Biliary complications: Complications in the biliary system with bilioma formation usually occur with ablation of tumors in the vicinity of biliary ducts and are more frequent with MW ablation techniques. The main bile ducts located near the hepatic hilum are considered protected from thermal damage by the heat-sink effect of the portal vein (with the exception of situations of diminished vascular flow). This is why it is difficult to completely ablate lesions located in the vicinity of the hepatic vascular hilum due to its cooling effect. The technique of choice for lesions in the vicinity of the hepatic hilum would be IRE, since it avoids the heat-sink effect and produces less damage to the biliary tract. On special occasions, cooling the biliary tract by draining it can be considered. Biliary strictures are one of the main complications, which when severe (jaundice, cholangitis or abscess) may require percutaneous biliary drainage.

Hepatic abscess: Hepatic abscess is one of the main major complications after tumor ablation (0.2-2%). The main causes are colonization of the biliary tract (bilioenteric anastomosis, endoscopic sphiterectomy, biliary stent or pneumobilia). Management of post-abdominal abscesses is usually the same as for any other etiology: small abscesses with antibiotic treatment is usually sufficient and large abscesses usually require percutaneous drainage.

2. Extrahepatic complications: Complications can be grouped as follows extrahepatic in 4 main groups:

- Penetration and thermal injury of adjacent organs.
- Implants in the needle pathway
- Thermal effect of adjacent organs.
- Pneumothorax - Hemothorax

Pneumothorax and Hemothorax: depend mainly on percutaneous access and needle trajectory, although sometimes it can be secondary to the treatment of lesions near the hepatic dome. It is advisable to perform a chest X-ray if the patient presents chest pain or dyspnea for confirmation. Both pneumothorax and hemothorax are usually self-limiting, although if the patient remains symptomatic a chest drainage tube may be necessary. Other less frequent complications include diaphragmatic hernias due to direct injury to the diaphragm.

Injury to hollow viscera: The treatment of hepatic subcapsular lesions can cause damage to the colon, small intestine and gastric chamber, the colon being very sensitive due to its thin wall and fixed position. Adhesions due to previous surgeries or chronic cholecystitis increase the risk of intestinal perforation due to decreased intestinal motility. A distance of at least 1 cm between the ablation area and the intestine is considered safe. Management of gastrointestinal tract (GIT) lesions includes antibiotherapy, drainage of infected fluid and abscesses, with surgery being the last option.

Cholecystitis: Acute cholecystitis may occur after ablation of masses adjacent to the gallbladder. It is possible to find minimal thickening of the gallbladder wall in subsequent controls, being exceptional cholecystitis or gallbladder perforation, since bile dissipates energy in the gallbladder fossa.

Tumor seeding: Tumor seeding in the needle track is extremely rare (0.3- 4%). The most important risk factors are: poorly differentiated tumor, subcapsular location, previous percutaneous biopsies, multiple treatment sessions, use of multiple needles. To avoid tumor seeding it is recommended to use the minimum number of punctures and repositioning attempts, to cross the shortest distance of healthy liver parenchyma and cauterization of the entry path after the procedure.

PATIENT SELECTION

The main limitation of thermal ablation (TA) has been local tumor progression (LTP) and tumor size. LTP is 3 times more frequent in tumors larger than 3 cm (8). The number of metastatic lesions is also important during patient selection, as patients with multiple tumors may have limited hepatic reserve and may have an increased risk of multifocal hepatic progression.Increasing knowledge of the factors affecting oncologic outcomes after ablation has allowed patients with small-volume resectable disease to be treated by AT with curative intent. Universally accepted indications for image-guided ablation include:

- Limited number of metastases (less than 5)

- Small tumors (up to 5 cm in greatest diameter)

- Patients who cannot undergo or refuse surgery. (9)

Potentially low-risk surgical candidates with a tumor size <3 cm and no extrahepatic disease, percutaneous TA with ablation margins greater than 10 mm may offer a chance of local cure similar to hepatectomy, avoiding surgical morbidity.(10)

Image-guided TA minimizes destruction of healthy liver tissue and is preferred to hepatectomy in patients with underlying cirrhosis or steatohepatitis as a result of prolonged exposure to chemotherapy, and in patients who have previously undergone extensive liver resection. Several uncontrolled studies of image-guided percutaneous TA reported overall survival outcomes comparable to hepatectomy (up to 55% at 5 years).(11)

A clinical risk score for ablation was proposed and indicated that certain patient and disease characteristics are more likely to predict oncologic outcomes after TA and local tumor progression after TA should be considered during patient selection (12).

There are a number of factors that are associated with decreased overall survival and local tumor progression-free survival (PFS) after TA such as (13):

- Tumor biology expressed as lymphovascular invasion at the time of primary resection.

- The disease-free interval from initial diagnosis to detection of liver metastases less than 12 months.

- Tumor size greater than 3 cm.

- CEA level greater than 30 ng/m (in the case of colon cancer metastases).

- The origin of the primary tumor, as primary tumors originating in the right colon are associated with worse oncologic outcomes compared to patients with left-sided CRC.(14)

PREOPERATIVE IMAGING

The importance of systematic preablation imaging for patient selection and procedure planning requires performing:

• A baseline computed tomography (CT) scan with intravenous contrast of the chest, abdomen and pelvis for the study of patients considered for ablation.

• Preoperative basal whole-body fluorodeoxyglucose (FDG) positron emission tomography (PET)/CT can provide additional information on hepatic and extrahepatic metastatic disease and may change treatment in a significant number of patients (15); also because we routinely use PET guidance during ablation, but also because metabolic imaging appears to be extremely sensitive for early detection of subsequent LTP and disease progression in general. (16)

• Magnetic resonance imaging (MRI) and, in particular, MRI with EOVIST contrast is the most anatomically accurate imaging modality for the detection and characterization of liver tumors, providing valuable information on patient eligibility for TA. It is especially useful for the detection of smaller tumors that may not be easily detected by CT and PET.

This allows visualization of the liver and tumor prior to contrast administration that resembles the appearance of the tumor on the day of ablation. Information on arterial supply and vascularity can also aid in planning the procedure.

EVALUATION OF THE ABLATION ZONE

All percutaneous TA techniques involve the insertion of an applicator (i.e., radiofrequency electrode, microwave antenna, cryoprobe or laser fiber) directly into or adjacent to the target tumor. Ultrasound (US), CT, MRI and PET can be used alone or in combination, for guidance. US can depict electrode insertion (targeting) and ablation zone (AZ) formation in real time, while PET and CT image fusion can provide additional guidance, as well as better assessment of the AZ and surrounding structures.(17)

The proximity of the tumor to critical structures is an important consideration for TA planning. Complications such as gastrointestinal perforation (seen in less than 0.2% of cases) can have a significant impact on morbidity. Hydro- or pneumodissection (i.e., instillation of fluid or air between structures) and balloon interposition techniques under US guidance or CT fluoroscopy may limit collateral injury to nearby structures, such as the GI tract, pancreas, kidney, or diaphragm.(18)

Post-ablation imaging evaluation should confirm complete tumor ablation with adequate ablation margins or detect residual unablated tumor. This evaluation should be performed by:

• A postablation contrast-enhanced computed tomography (CECT) scan is most commonly used to provide a rapid assessment of the ablated area, depicted as a low-attenuation unenhanced area. Post-ablation venous phase CT images can be fused with pre-procedural CT, PET or MRI images to delineate the ablation zone (19) CT evaluation of the ablated lesion should be performed shortly and ideally immediately after tumor ablation.

• An immediate postablation contrast-enhanced ultrasound followed by CECT to evaluate the AZ and confirm complete coverage of the target tumor (20).

• A dynamic CT or MRI scan 3-8 weeks after TA is mandatory to confirm complete ablation of the target tumor.

The first postablation image will serve as the new reference image for further evaluation of AZ and detection of local tumor progression.

BIBLIOGRAPHY .

1. Salati U, Barry A, Chou FY, Ma R, Liu DM. State of the ablation nation: a review of ablative therapies for cure in the treatment of hepatocellular carcinoma. Future Oncol 2017; 13:1437-1448.

2. Wells SA, Hinshaw JL, Lubner MG, Ziemlewicz TJ, Brace CL, Lee FT Jr. Liver ablation: best practice. Radiol Clin North Am 2015; 53:933-971.

3. Seror O. Ablative therapies: Advantages and disadvantages of radiofrequency, cryotherapy, microwave and electroporation methods, or how to choose the right method for an individual patient? Diagnostic and Interventional Imaging. 2015; 96 (6):617-24

4. Goldberg SN, Gazelle GS, Mueller PR. Thermal ablation therapy for focal malignancy: A unified approach to underlying principles, techniques, and diagnostic imaging guidance. Am J Roentgenol. 2000; 174(2):323-31.

5. Ahmed M. Image-guided tumor ablation: standardization of terminology and reporting criteria- a 10-year update. Radiology. 2014; 273(1):241-60.

6. The Research Group on Ablation Therapies in Oncologic Surgery (METABLATE). Kouri BE, Abrams RA, Al-Refaie WB, Azad N, Farrell J, Gaba RC, et al. ACR appropriateness criteria radiologic management of hepatic malignancy. J Am Coll Radiol. 2016; 13(3):265- 73.

7. Lahat E, et al. Complications after percutaneous ablation of liver tumors: a systematic review. Hepatobiliary Surg Nutr 2014;3(5):317- 323.

8. Tanis E, Nordlinger B, Mauer M, et al: Local recurrence rates after radiofrequency ablation or resection of colorectal liver metastases. Analysis of the European Organisation for Research and Treatment of Cancer #40004 and #40983. Eur J Cancer 2014; 50: 912-919.

9. Gillams A, Goldberg N, Ahmed M, et al: Thermal ablation of colorectal liver metastases: a position paper by an international panel of ablation experts, the interventional oncology sans frontières meeting 2013. Eur Radiol 2015; 25: 3438-3454.

10. Shady W, Petre EN, Do KG, et al: Percutaneous microwave versus radiofrequency ablation of colorectal liver metastases: Ablation with clear margins (A0) provides the best local tumor control. J Vasc Int Radiol 2018; 29: 268-275.

11. Meijerink MR, Puijk RS, van Tilborg AAJM, et al: Radiofrequency and

microwave ablation compared to systemic chemotherapy and to partial hepatectomy in the treatment of colorectal liver metastases: A systematic review and meta-analysis. Cardiovasc Int Radiol 2018; 41: 1189-1204.

12. Shady W, Petre EN, Gonen M, et al: Percutaneous Radiofrequency ablation of colorectal cancer liver metastases: Factors affecting outcomes a 10-year experience at a single center. Radiology 2016; 278: 601-611.

13. Gu Y, Huang Z, Gu H, et al: Does the Site of the Primary Affect outcomes when ablating colorectal liver metastases with radiofrequency ablation? Cardiovasc Int Radiol 2018; 41: 912-919.

14. Yamashita S, Odisio B, Huang S, et al: Embryonic origin of primary colon cancer predicts survival in patients undergoing ablation for colorectal liver metastases. HPB 2019; 21: S584-S5S5.

15. Kishore SA, Drabkin MJ, Sofocleous CT: Fluorodeoxyglucose-PET for ablation treatment planning, intraprocedural monitoring, and response. PET Clinics 2019; 14: 427-436.

16. Cornelis FH, Petre EN, Vakiani E, et al: Immediate Postablation18F- FDG injection and corresponding SUV are surrogate biomarkers of local tumor progression after thermal ablation of colorectal carcinoma liver metastases. J Nucl Med 2018; 59: 1360-1365.

17. Ahmed M, Solbiati L, Brace CL, et al: Image-guided tumor ablation: standardization of terminology and reporting criteria-A 10-year update. Radiology 2014; 273: 241-260.

18. Garnon J, Cazzato RL, Caudrelier J, et al: Adjunctive Thermoprotection during percutaneous thermal ablation procedures: Review of current techniques. Cardiovasc Int Radiol 2018; 42: 344-357.

19. Solbiati M, Muglia R, Goldberg SN, et al: A novel software platform for volumetric assessment of ablation completeness. Int J Hyperther 2019; 36: 336-342.

20. Mauri G, Porazzi E, Cova L, et al: Intraprocedural contrast-enhanced ultrasound (CEUS) in liver percutaneous radiofrequency ablation: clinical impact and health technology assessment. Insights Imaging 2014; 5: 209-216.

INTRODUCTION

Intraoperative radiofrequency ablation (IRFA) of liver metastases was first reported by Curley et al (1) in 1999 and Elias et al (2) in 2000.

RFA represents the oldest and most studied energy-based percutaneous ablation modality to date. This technique uses high-frequency alternating electric current, in the radiofrequency range of 200 to 1200 MHz, which is delivered to the tumor area by means of a needle electrode placed in the target tissue either percutaneously or laparoscopically. Ablation can be performed in different situations:

- As the only therapeutic option.
- Combined with surgical treatment.
- As initial treatment and subsequent definitive therapy
- Associated with neoadjuvant chemotherapy treatment. That prepares it for subsequent treatment (liver transplantation).

The size of the lesions does not currently determine radiofrequency treatment. However, the efficacy of the treatment decreases notably when treating lesions with a diameter greater than 8 cm, even if they are treated in two or three sessions.

It is essential to locate the hepatic lesions with preferably ultrasound control in order to be able to perform the ablative treatment. Small lesions located especially in the posterior superior part of the right lobe are difficult to access, given the proximity of the diaphragm and lung, and require for their treatment a wide experience in hepatic punctures.

The number of lesions may condition the need to perform the treatment in several sessions.

MECHANISM OF ACTION

RFA causes cell death by thermocoagulation necrosis. The mechanism of cell death in RFA is based on the dissipation of electrical energy as frictional heat (Fig. 1). Therefore, the effectiveness of RFA depends on tissue conductivity, which is strongly correlated with water content. Two types of RFA devices are used clinically:

- Monopolar RFA (MP) uses a single antenna. Intratumoral electrodes are inserted centered on the area to be destroyed. For metastases smaller than 3.0 cm, the monopolar mode reliably provides reproducible spherical zones of necrosis. For larger lesions, multifocal or multiple overlapping applications of RFA may be necessary (3).

- Bipolar RFA (BP) uses dual antennas, or two electrodes on the same antenna facing each other, resulting in ablation of a much larger tumor volume compared to MP in a single ablation. It is less affected by the heat sink effect compared to MP

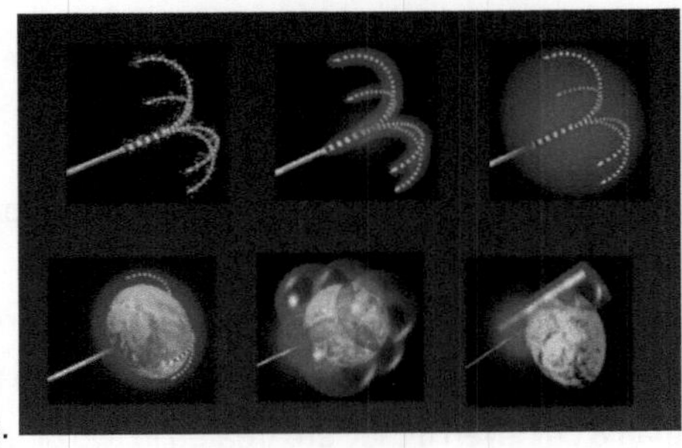

(4).

Fig. 1: device in radiofrequency technique.

According to the size and shape of the needle tip, a spherical ablated area generally 2 to 5 cm in diameter is generated in about 10 to 30 minutes. The active tissue heating zone is limited to a few millimeters surrounding the active electrode, and the rest of the ablation zone is heated by thermal conduction. With an increase in lesion size, treatment efficacy decreases, as the maximum result is obtained for volumes less than 3.5 cm. RFA is limited by increased impedance and excessive local temperature. Several technical devices are available to avoid this effect, such as temperature or impedance control during the procedure or simultaneous instillation of saline solution into the tissue surrounding the RF needle (5, 6,7). Temperatures above 100°C lead to dehydration and charring of the tissue, limits the heating capacity of an RFA probe and also makes the RFA particularly sensitive to the heat dissipating effect of blood flowing in adjacent vessels.

RFA MODALITIES AND UTILIZATION PATTERNS

1. _Modalities of RFA_: RFA can be applied in different ways:

- **Percutaneous route**: Percutaneous RFA is generally used to treat patients with liver metastases who are at high risk for surgery. Percutaneous ultrasound-guided percutaneous RFA (Fig. 2) enables a

The most frequently used is a quick and inexpensive outpatient treatment with relatively easy access.

- **Laparoscopic surgery:** The minimally invasive laparoscopic approach can be performed on an outpatient basis. It has most of the advantages of an open approach, although with some limitations due to the technical difficulties of laparoscopic liver mobilization and the use of ultrasound. Its main indication is in peripheral lesions where the risk of bleeding is higher or when further exploration of the abdominal cavity or liver parenchyma is required in patients without previous surgery.

- **Open surgery**: The benefits of a surgical approach to RFA are: optimal intra-abdominal staging (detection of unsuspected carcinomatosis and increased diagnostic sensitivity of intra-operative ultrasound), the ability to perform simultaneous resection, perform the Pringle maneuver and better protect adjacent organs. It is used when associated with major liver resection surgery or when surgical liberation of the liver from anatomical structures intimately adhered to it is required in patients who have previously undergone surgery.

Fig. 2: Ultrasound guided radiofrequency ablation

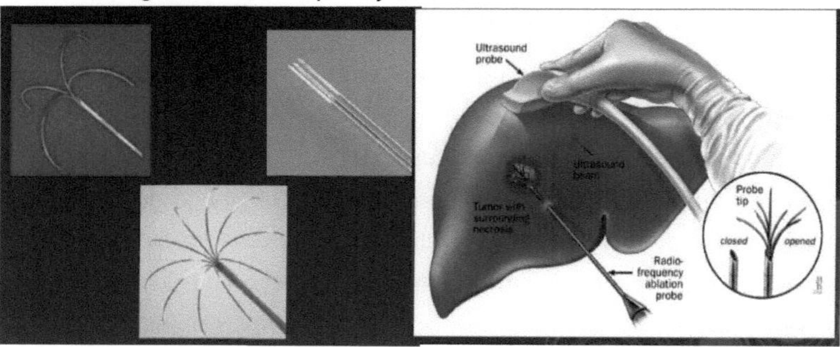

A number of advantages and disadvantages of each approach have been detailed and are summarized in Table 1 (8, 9).

Table 1: Advantages and disadvantages of approach routes for RFA

	Advantages	**Disadvantages**
Percutaneous route	Allows fast and cost-effective outpatient treatment with relative ease of accessibility	Inability to accurately assess the adequacy of the thermal injury zone.
		Lack of reliability when used in certain locations
	Use of local anesthesia	Intraoperative ultrasound has worse tumor staging
Open road	Perform maneuver Pringle maneuver	Invasive technique
	Combines RFA + liver resection	>indexhospitalization and recovery
		Need for general anesthesia
		Cost increase
limitations	Best identification of tumor margins and satellite nodules	Difficult technique for deep tumors and in segments VI, VII and VIII.
	Best results of complete coagulation	

2. *The utilization guidelines* for RFA are:

- The number of lesions should not be considered an absolute contraindication for RFA, but most centers preferentially treat patients with ≤5 lesions.

- The best complete ablation rates are achieved in lesions with a maximum diameter of ≤3 cm.

- Tumor location can be a problem in case of:

• injury to the surface of the liver due to thermal injury to adjacent structures (although this risk can be mitigated by percutaneous pre-ablation infusion of dextrose fluid into the space between the liver and adjacent bowel, abdominal wall or diaphragm)

• lesions adjacent to the hepatic hilum (risk of thermal injury to the biliary tract)

• lesions adjacent to large hepatic vessels (due to the heat sink phenomenon)

INDICATIONS AND CONTRAINDICATIONS OF RFA

1. **Indications** (table 2)

a. Open RF is indicated:

- when there is a curative intention
- in tumors located very deep in the hepatic parenchyma and inaccessible by laparoscopic approach
- in the patient with synchronous metastases as it can be combined with resection of the primary tumor (10).

b. Percutaneous RF is indicated:

- in patients at higher risk for RF by open laparoscopy
- when performed for palliative purposes, to improve local control, prevent growth or prolong life
- in patients with recurrence after RF or with progressive lesions (11);
- in patients who refuse liver resection.

c. Laparoscopic RF is indicated when there is curative intent. It is a less invasive method, with a shorter hospital stay, an early recovery of the patient and less postoperative pain than with open RF. Laparoscopic intraoperative ultrasound allows better tumor staging than percutaneous ultrasound, diagnoses 30-38% of liver lesions not visualized in the preoperative period, which improves the results of RF (12). With laparoscopy, extrahepatic disease previously undiagnosed by imaging tests is observed in 12% of the cases.

Table 2: RFA indications

RFA PERCUTANEOUS	*RFA OPEN SURGERY*	*LAPAROSCOPIC RFA*
In palliative treatment	Healing intent	Healing intent
Relapse after RF	Intraparenchymal tumors deep	Peripheral metastases accessible
Rejection of section hepatic	Synchronous metastases	
Increased risk for RF with open surgery		

2. General *limitations* and **contraindications** of RFA are shown in Table 3. RFA is an effective, safe and widely available therapy (13). The main limitations are:

- the high frequency of local recurrence, mainly for lesions larger than 3 cm (14).
- possible incomplete ablation of the lesion near large vessels (>3 mm in diameter). In this context, multiple overlapping probe placements are required to achieve larger ablation zones, making the procedure more consuming and less safe for the patient (14).

Table 3: Limitations and contraindications of radiofrequency ablation

Limitations	Contraindications
Tumors>3cmnoconsiguenecrosis complete tumor	Dilation fro m ducts biliary intrahepatic
Difficult access to tumors in proximity to vascular structures	Intractable coagulopathy
Difficult access to tumors in certain segments	bilioenteric anastomoses
Subcapsular lesions may fragment into the peritoneum.	
Damage to the spinal cord by ablation of . segment IVb and V	Tumors located < 1 cm of the bile duct. the risk of late stenosis of the main bile duct.
Management of patients with cirrhosis is more complicated.	
Satellite nodules and infiltrative growth pattern	
Variability in electrical and thermal conductivity	

HIGH RISK LOCATIONS

1. Adjacent to great vessels: < 5mm from a primary or secondary branch of the portal vein, base of the suprahepatic veins or IVC.
2. Adjacent to extrahepatic organs: < 5mm from cornea, lung, gallbladder, right kidney or GI tract.

COMPLICATIONS

Complications are low and can be divided into early and late (more than 30 days). They are associated with the number of sessions performed and not with tumor size or the type of equipment used. Mulier et al described that complications with percutaneous, laparoscopic or laparotomy radiofrequency were 7.2%, 9.5% and 9.9% respectively. Risk factors identified for complications include (15):

- Bilirubin > 2.0 mmol / dL
- Cirrhosis
- Subcapsular or deep central tumor location,
- Multiple tumors requiring multiple RFA applications
- Presence of a biliary-enteric anastomosis.
- Surgeon experience of < 50 procedures.

Some complications are specific to RFA and, therefore, can also be induced by IRFA. Some complications occur more frequently with percutaneous RFA than with the open surgical approach. The most frequently reported complications are:

- Bleeding (1.6%):
- Abscess (1.1%): They appear after an asymptomatic period ranging from 8 days to 5 months. The presence of a low amount of gas in the necrotic area during the month following RFA is common. Diabetes mellitus, bilioenteric anastomosis and a history of sphincterotomy, biliary stent, biliary stenosis, etc., are factors in the development of RFA.

risk for liver abscess after RFA ablation procedures (16).

- Biliary ductal injury (1%) and strictures: Biliary strictures appear after fibrous healing of thermal damage to the biliary tract. Many of these biliary strictures are asymptomatic and do not require treatment. Biliary strictures that occur near the main biliary tract after RFA are symptomatic and can be treated with intrahepatic stenting or endoscopic sphincterotomy.(17)
- Malignant seeding of the puncturing tract (0.9%): Indirect punctures through an area of healthy liver and coagulation of the tract are recommended to reduce the risk of bleeding and tumor implantation, especially for subcapsular lesions (18).
- gastrointestinal perforation (0.3%),

- pneumothorax and/or pleural effusion

- Acute cholecystitis: Cholecystitis due to thermal injury is exceptional and is asymptomatic in most cases. The risk of gallbladder perforation is present only when large tumors are adjacent to the gallbladder.

- Vascular thrombosis: caused by thermal endothelial damage. Vascular occlusion increases the risk of vascular thrombosis, especially when the distance is < 5 mm. Portal thromboses are more frequent in cirrhotic patients and may be complicated by liver failure. Thrombosis of veins <3 mm in diameter is common after IRFA, are asymptomatic and spontaneous regression is often observed after 2 months. Hyperthermia-related vascular thrombosis is only seen in vessels smaller than 3 mm and are asymptomatic. Larger caliber vessels are protected by the continuous cooling effect of the blood flow.(19)

- Skin burns are rare since the introduction of modern systems using grounding plates with a much larger surface area. Skin burns occur when the scattering surface is inadequate for the radiofrequency power.

EVALUATION OF THE THERAPEUTIC EFFECT.

The success of the ARF depends on several parameters:

- the ability to detect and target all lesions in need of treatment
- ability to monitor in real time the range
- The effectiveness of the treatment (ultrasound, computed tomography or magnetic resonance imaging).

Local recurrence in liver tumors smaller than 3 cm after radiofrequency treatment by laparoscopy and/or laparotomy is 3.6% compared to percutaneous which is 16%. In tumors larger than 5 cm in percutaneous treatment it is up to 60% and in laparotomy and/or laparoscopy it is between 40-50%. The two factors that most influence post-radiofrequency tumor recurrence are:

- Tumor size: Tumors smaller than 3 cm have a better prognosis.
- Type of treatment approach: the laparoscopic and/or laparotomic approach has a better prognosis, provides better tumor control and diagnosis of previously unknown occult liver lesions by intraoperative ultrasound.

Local tumor recurrence rates ranged from 6% to 40% and were related to the size, number and location of the lesions (21). RFA has a mortality of 0 to 2% and the most frequently reported morbidity rate is 6 to 9.

BIBLIOGRAPHY .

1. Curley SA, Izzo F, Delrio P, Ellis LM, Granchi J, Vallone P, et al. Ablation par radiofréquence des tumeurs malignes hépatiques primaires et métastatiques non résécables: résultats chez 123 patients. Ann Surg. 1999; 230:1-8.

2. Elias D, Goharin A, El OA, Taieb J, Duvillard P, Lasser P , et al. Usefulness of intraoperative radiofrequency thermoablation of liver tumors associated or not with hepatectomy. Eur J Surg Oncol. 2000; 26
:763-76

3. Goldberg SN, Gazelle GS, Mueller PR. Thermal Ablation Therapy for Focal Malignancies: A Unified Approach to Underlying Principles, Techniques, and Imaging Guidance. AJR Am J Roentgenol 2000; 174:323-331.

4. Strap CL. Radiofrequency and microwave ablation of the liver, lung, kidney and bone: what are the differences? Curr Probl Diag Radiol 2009; 38:135-143.

5. Crocetti L, de Baere T, Lencioni R. Quality improvement guidelines for radiofrequency ablation of liver tumours. Cardiovasc Intervent Radiol 2010; 33:11-17.

6. Shimizu A, Ishizaka H, Awata S et al. Enlargement of the volume of radiofrequency ablation by injection of saturated saline solution of NaCl into the vaporization zone. Acta Radiol 2009; 50: 61-64.

7. Cirocchi R, Trastulli S, Boselli C, Montedori A, Cavaliere D, Parisi A, et al: Radiofrequency ablation in the treatment of liver metastases from colorectal cancer. Cochrane Database Syst Rev 2012; 6: CD006317.

8. Iwai S, Sakaguchi H, Fujii H, Kobayashi S, Morikawa H, Enomoto M, et al: Benefits of artificially induced pleural effusion and/or ascites for percutaneous radiofrequency ablation of hepatocellular carcinoma located on the liver surface and in the hepatic dome. Hepatogastroenterology 2012; 59: 546-550.

9. Dodd G.D., Napier D., Schoolfield J.D. and Hubbard L.: Percutaneous radiofrequency ablation of hepatic tumors: postablation syndrome. AJR Am J Roentgenol 2005; 185: 51-57.

10. Machi F, Uchida S, Sumida K, Limm WM, Hundahl SA, Oishi AJ, et al. Ultrasound-guided radiofrequency thermal ablation of liver tumors percutaneous, laparoscopic, and open surgical approaches. J Gastrointestinal Surg 2001; 5: 477-89.

11. Wood TF, Rose DM, Cheng M, Allegra DP, Foshag LJ, Bilchik AJ. Radiofrequency ablation of 231 unresectable hepatic tumors: Indications, limitations, and complications. Ann Surg Oncol 2000; 7: 593-600.

12. Smith MK, Mutter D, Forbes LE, Mulier S, Marescaux J. The physiologic effect of the pneumoperitoneum on radiofrequency ablation. Surg Endosc 2004; 18: 35-8.

13. Van Cutsem E, Cervantes A, Adam R et al. ESMO consensus guidelines for the treatment of patients with metastatic colorectal cancer. Ann Oncol 2016; 27:1386-13422.

14. Kim SK, Rhim H, Kim YS et al. Radiofrequency thermal ablation of liver tumors: pitfalls and challenges. Images of the abdomen. 2005; 30:727- 733.

15. Zagoria RJ, Chen MY, Shen P, Levine EA. Complications of radiofrequency ablation of liver metastases. I'm Surgery. 2002; 68: 204- 209.

16. Elias D, Di PD, Gachot B, Menegon P, Hakime A, De Baere T. Liver abscess after radiofrequency ablation of tumors in patients with a biliary tract procedure.Gastroenterol Clin Biol. 2006; 30:823-827.

17. Kim SH, Lim HK, Choi D, Lee WJ, Kim SH, Kim MJ, et al. Bile duct changes after radiofrequency ablation of hepatocellular carcinoma: frequency and clinical significance.   AJR Am J Roentgenol. 2004; 183:1611-1617.

18. Livraghi T, Lazzaroni S, Meloni F, Solbiati L: Risk of tumour seeding after percutaneous radiofrequency ablation for hepatocellular carcinoma. Br J Surg 2005; 92: 856-858.

19. Lu D.S., Raman S.S., Vodopich D.J., Wang M., Sayre J., Lassman C.: Effect of vessel size on creation of hepatic radiofrequency lesions in pigs: assessment of the "heat sink" effect. AJR Am J Roentgenol 2002; 178: 47-51.

20. Leyendecker J.R., Dodd G.D., Halff G.A., McCoy V.A., Napier D.H., Hubbard L.G., et al: Sonographically observed echogenic response during intraoperative radiofrequency ablation of cirrhotic livers: pathologic correlation. AJR Am J Roentgenol 2002; 178: 1147-1151.

21. Wong SL, Mangu PB, Choti MA. American Society of Clinical Oncology 2009 review of clinical evidence on radiofrequency ablation of liver metastases from colorectal cancer. J Clin Oncol. 2010; 28:493-508.

Chapter 3: Microwave Ablation

INTRODUCTION AND TECHNIQUE

Microwave ablation is an ablative technique subsequent to radiofrequency (1). The technique is based on the use of electromagnetic waves. The electromagnetic field generates homogeneous heat destruction causing necrosis by coagulation of the tumor tissue (2).

The initial technique, which consisted of a microwave-generating antenna, has been refined since 2007 to the present time, so that there are devices with longer applicators and multiple antennas, as well as a cooling system for these antennas. Among the improvements in the latest generation of MWA is a shortening of the ablation time and the generation of an intense field of almost spherical shape that improves the extent of tissue coagulation (Fig. 1).

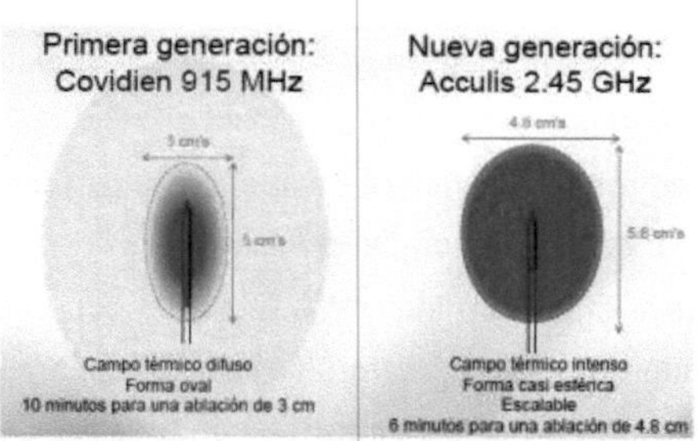

Fig. 1: Ablation improvements of the latest models

The goal of thermal ablation is to heat target tissues to temperatures that can induce immediate coagulative necrosis (typically greater than 60 °C). A complete treatment should cover the tumor plus a 5-10 mm safety margin (analogous to a surgical margin) without affecting healthy parenchyma and vulnerable non-target structures (3, 4,5). In MW ablation, the mechanism of heat generation is based on the rapid frictional motion of water molecules in the high-frequency electromagnetic field. Microwaves are able to effectively heat and propagate through many types of tissue, even those with low electrical conductivity, high impedance or low thermal conductivity. There are currently 3 generations of commercially available devices consisting of: 1) a programmable electromagnetic wave generator, 2) a disposable applicator(s) for direct release of energy into the patient's body, 3) an interstitial applicator for use via percutaneous or intraoperative, 4) a flexible applicator (intracavitary use) and 5) a peristaltic pump for forced circulation of fluid inside the applicator in use for cooling purposes, the third generation features built-in antenna cooling systems and high-powered generators (6,7). (Fig.2)

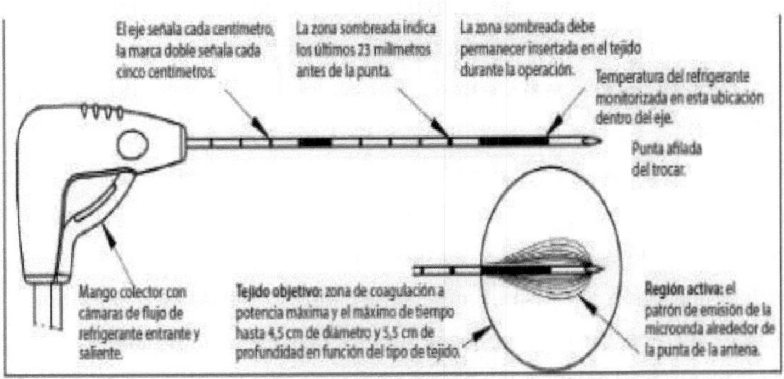

Fig. 2: Application device with cooling system
Microwave radiation is generated by electromagnetic waves at a certain frequency, from 900 to 2450 MHz. This procedure achieves tumor reduction and/or complete ablation (8). Among the improvements in the latest generation AMO is a shortening of the ablation time and the generation of an intense field of almost spherical shape that improves the extent of tissue coagulation.

There are different ways of performing MWA (percutaneous, laparoscopic or open surgery). The percutaneous approach offers several advantages (3,9), it is the least invasive, relatively expensive, can be performed on an outpatient basis and can be repeated to treat recurrent tumors. All patients should undergo ultrasound (US) which is the most useful, contrast-enhanced ultrasound (CEUS) and contrast-enhanced computed tomography (CT) or gadolinium–enhanced magnetic resonance imaging (MRI) to delineate the target tumor prior to MWA (10).

To destroy possible micrometastases or microscopic foci around the tumor and to prevent local tumor progression, ablation with a 5-10 mm margin around the tumor has been recommended (11). In many cases, multiple overlapping ablations may be necessary to create the required ablation zone size. Overlapping ablations can be created in 2 ways: a) multiple insertions of a single antenna to create sequentially overlapping ablation zones or b) multiple antennas ablating simultaneously to create a confluent ablation zone.

INDICATIONS AND CONTRAINDICATIONS

1. *Indications:* the Clinical Guidelines on microwave ablation of liver tumors recommend its use both for curative and palliative purposes. The indications for curative treatment coincide with Milan's criteria (12). The indication for AMO is referred to patients with these characteristics who have contraindicated surgery, which is indicated in this situation by the European Guidelines (EASL-European Association for the Study of the Liver (13). The current indications for tumor ablation are also expanded to intrahepatic cholangiocarcinoma (14) (Table 1).

Table 1: Indications for microwave ablation treatment

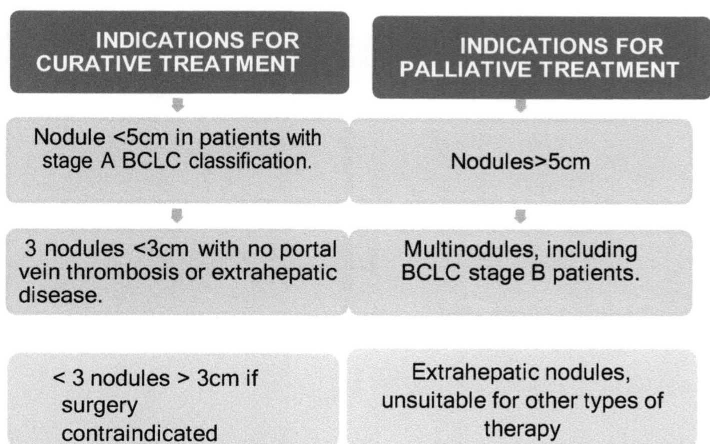

2. *Contraindications* include its use in patients with pacemakers or other electronic implants, which could be due to the high temperatures reached during ablations (above 100ºC) and transmitted to adjacent areas (Fig.3).

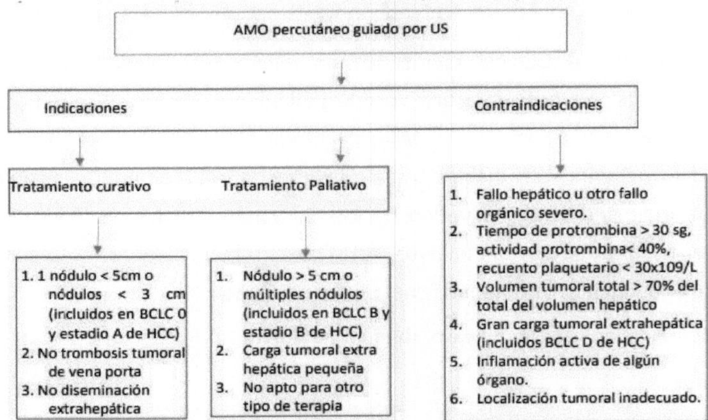

Fig. 3: Indications and contraindications of AMO

COMPLICATIONS

Microwave ablation produces major and minor complications, which are described in Table 1.

Table 1: Complications of microwave ablation therapy

Major complications	Minor complications
Intraperitoneal bleeding	**Immediate:**
Portal vein thrombosis	Pain
Intrahepatic hematoma	Post-ablation syndrome
Biliary leakage	Skin burn
Bilioma	
Bile duct injury / biliary stenosis	**Peri procedure:**
Hepatic dysfunction	Fever
Hepatic abscess	Asymptomatic pleural effusion
Intestinal perforation	Vesicular wall thinning
Diaphragmatic hernia	Arterio-portal shunt asymptomatic
Hemothorax / Refractory pleural effusion	
Tumor seeding	**Late:**
Burns	Bile duct contraction
Empyema	

In recent years, a comparison has been made between radiofrequency ablation and microwave ablation in all aspects (15,16). Table 2 shows the similarities and differences of each technique.

Table 2: comparison of radiofrequency and microwave ablation methods

ARF	AMO
• Electric current • Grounding pad required (risk of burns) • Tissue carbonization and boiling increase impedance and reduce electrical energy and conductivity. impedance and reduce electrical energy and conductivity. • Lower intratumoral temperature • More pain after the procedure • Non-predictable ablation zone • Sink effect • Treatment of solitary nodules and multiple lesions • Longer duration of sessions • Lower ablation volume • Similar complication rate • Contraindicated metal clips and pacemakers	• Electromagnetic energy • No grounding pad required (risk of burns due to high temperature) • Fast and homogeneous tissue heating • Ionic polarization • High intratumoral temperature • Reduced post-procedure pain • Predictable ablation zone • Reduced susceptibility to sink effect • Multimultaneous treatments of multiple injuries • Shorter duration of sessions • Increased ablation volume • Initially, surgical clips would not be contraindicated, but patients with pacemakers and pacemakers are contraindicated. electrical implants.

BIBLIOGRAPHY .

1. K T. A new operative procedure for hepatic surgery using a microwave tissue coagulator. Nippon Geka Hokan. 1979; 48:160-72.

2. AR G. Image guided tumor ablation. Cancer Imaging. 2005; 5:103-9.

3. Vogl TJ, Nour-Eldin NA, Hammerstingl RM, Panahi B, Naguib NNN. Microwave ablation (MWA): basics, technique and results in primary and metastatic liver neoplasms. Rofo 2017; 189:1055-1066.

4. Meloni MF, Chiang J, Laeseke PF, et al. Microwave ablation in primary and secondary liver tumours: technical and clinical approaches. Int J Hyperthermia 2017; 33:15-24.

5. Seror O. Ablative therapies: Advantages and disadvantages of radiofrequency, cryotherapy, microwave and electroporation methods, or how to choose the right method for an individual patient? Diagn Interv Imaging 2015; 96:617-624.

6. Lubner MG, Brace CL, JL H, [et al]. Microwave Tumor Ablation: Mechanism of Action, Clinical Results, and Devices. J Vasc Interv Radiol. 2010; 21: S192-S203.

7. Imajo K, Ogawa Y, Yoneda M, Saito S, Nakajima A. A review of conventional and newer generation microwave ablation systems for hepatocellular carcinoma. J Med Ultrason 2020; 47:265-277.

8. Poulou LS, Botsa E, Thanou I, Ziakas PD, Thanos L. Percutaneous microwave ablation vs radiofrequency ablation in the treatment of hepatocellular carcinoma. World J Hepatol. 2015; 7(8):1054-63.

9. Baker EH, Thompson K, McKillop IH, et al. Operative microwave ablation for hepatocellular carcinoma: a single center retrospective review of 219 patients. J Gastrointest Oncol 2017; 8:337-346.

10. Kurumi Y, Tani T, Naka S, et al. MR-guided microwave ablation for malignancies. Int J Clin Oncol 2007; 12:85-93.

11. Patel PA, Ingram L, Wilson ID, Breen DJ. No-touch wedge ablation technique of microwave ablation for the treatment of subcapsular tumors in the liver. J Vasc Interv Radiol 2013; 24:1257-1262.

12. Starr MKZ, Milan. Multiple criteria decision making, Amsterdam, North Holland1977.

13. European Association for the Study of the L, European Organisation for R, Treatment Of C. EASL-EORTC clinical practice guidelines: management of

hepatocellular carcinoma. J Hepatol. 2012; 56(4):908- 43.

14. Zhang K, Yu J, Yu X, et al. Clinical and survival outcomes of percutaneous microwave ablation for intrahepatic cholangiocarcinoma. Int J Hyperthermia 2018; 34:292-297.

15. Seror O. Ablative therapies: Advantages and disadvantages of radiofrequency, cryotherapy, microwave and electroporation methods, or how to choose the right method for an individual patient? Diagnostic and Interventional Imaging. 2015; 96 (6):617-24.

16. Yu J, Liang P, Yu X, Liu F, Chen L, Y W. A comparison of microwave ablation and bipolar radiofrequency ablation both with an internally cooled probe: results in ex vivo and in vivo porcine livers. Eur J Radiol. 2011; 79:124-30.

Chapter 4: Ablation by electroporation irreversible

INTRODUCTION

IRE is a newer ablation technology compared to RFA or MWA; which, unlike both, is non-thermal. By delivering direct high-voltage, low-energy electrical pulses to tumor cells, IRE triggers apoptosis and leads to controlled cell death.

The term electroporation is a phenomenon that has been known for several decades and refers to the increase in the permeability of the cell membrane by means of high magnitude electric fields. These fields are capable of altering the resting potential of the cell membrane in such a way that the structure of the lipid bilayer becomes unbalanced, giving rise to pores. When the pulses are of low magnitude, the process is reversible, as the cell is able to repair these defects and can continue living; however, with high electric fields, the cell cannot repair the defects and this leads to cell death (1).

Technological advances in recent decades and the rapid development of different image-guided ablation techniques have greatly expanded the therapeutic possibilities for surgically incurable liver metastases. For metastases that are not amenable to resection, nor to thermal ablation with radiofrequency or microwave ablation due to their proximity to blood vessels or bile ducts, irreversible electroporation (IRE) is increasingly used (2).

TECHNIQUE AND MECHANISM OF ACTION

1. ***Technique***: Irreversible electroporation is a percutaneous or, less frequently, laparoscopic procedure requiring general anesthesia and neuromuscular blocking agents; the patient is placed in the supine or left lateral decubitus position. The important thing is to avoid involuntary muscle contraction that may arise accidentally fom procedure-induced electrical stimulation of motor neurons. The exact measurements of the target lesion are assessed by US or CT, which determines the number and configuration of the electrodes (Fig. 1). Today, mainly two to six parallel needle electrodes (0-1 mm) are used. In hepatic IRE, the length of the electrodes is set at 20 mm (FIG. 2). Fifty to 100 electrical pulses are administered sequentially. To mitigate the risk of arrhythmia, the IRE is synchronized with the ECG with the absolute refractory period of the myocardial cells (3); the electrical pulse must be applied exactly 50µs after the R wave to coincide with the absolute refractory period of the myocardium in the cardiac cycle, with patients being under general anesthesia and under effect of muscle plaque relaxants such as rocuronium to avoid muscle contractions that trigger the pulses. IRE produces a strong electric field to permanently deactivate target cell homeostasis and induce cell death; an electric field of 1000-1500 V/cm is mandatory (4).

Fig. 1: *Definition of depth, length and width of the treatment area in relation to the electrode application.*

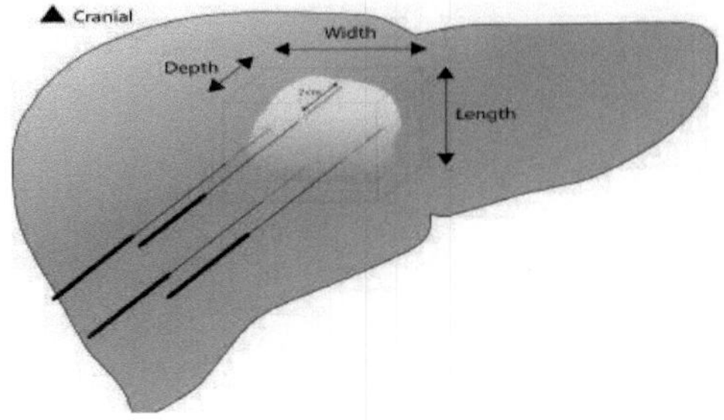

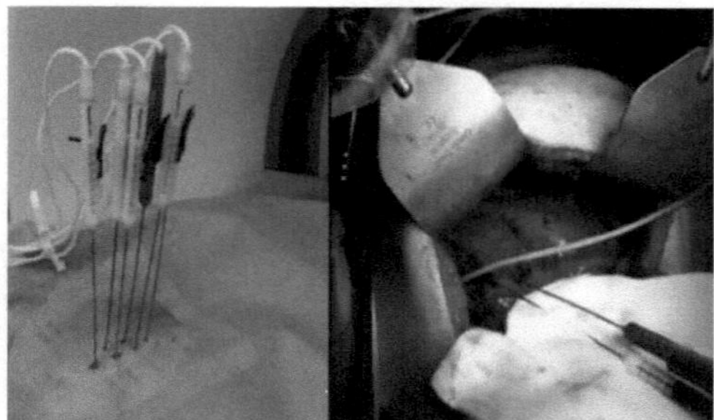

Fig. 2: CT-guided (left) and open (right) IRE.

The electrodes are placed parallel to each other (maximum angulation of 10°) to promote homogeneous energy delivery. The ablation zone should completely cover the tumor and a tumor-free margin of at least 5 mm in all directions (5). Because the calculated ablation zone extends at least 5 mm outward from the electrodes, they should be placed at the outer edge or just adjacent to the tumor (6) (Fig. 3).

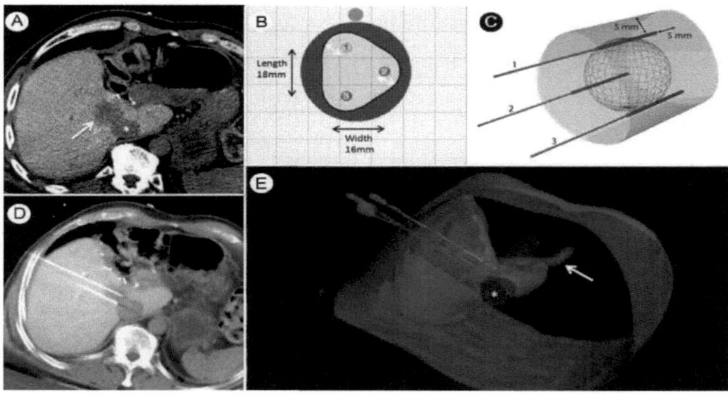

Fig. 3: (A) Pre-IRE CT shows a central hypoattenuating lesion (arrow). The asterisk represents the inferior vena cava. (B) Planning of the electrode configuration, with the yellow circle representing the tumor, the white arrows represent the expected tumor-free margin. (C) Calculated ablation zone extending 5 mm outward from each electrode in all directions. (D) CT fluoroscopy with 2 of 3 electrodes placed at the tumor periphery. (E) Three-dimensional reconstruction of the electrodes placed at the tumor periphery and the proximity of the tumor to the inferior vena cava (asterisk) and common bile duct (arrow).

2. **Mechanism of action:** The IRE system is made in the form of needles in monopolar or bipolar application and is named NanoKnife® (Angiodinamics, NY, USA) (Fig. 4).

The mechanism of death can range from apoptosis to coagulation necrosis, but is usually not a mechanism dependent on elevation of tissue temperature. The latter variety, known as irreversible electroporation (IRE), is especially useful because of the emergence of more powerful generators.(7)

The mechanism of operation of IRE is based on electrical energy; high voltage electrical pulses cause irreversible disruption of the cell membrane, leading to cell death; while the underlying connective tissue remains intact (8). Although heat development is an unavoidable side effect of electrical pulses, it does not

This temperature increase is thought to be detrimental to the surrounding connective tissue. Therefore, encrustation of vulnerable structures such as bile ducts and blood vessels remain apparent.(9)

The mechanism of action is to induce pores in the cell membrane that lead to cell apoptosis; it has demonstrated its capacity to destroy tissue without presenting the undesirable effects of thermal ablation such as the "heat sink effect" that occurs in the proximity of blood vessels. It also maintains the integrity of the extracellular matrix, so that structures such as blood vessels and bile ducts are not affected by irreversible electroporation (Fig. 6). Its use has been expanding in recent years in liver ablation.

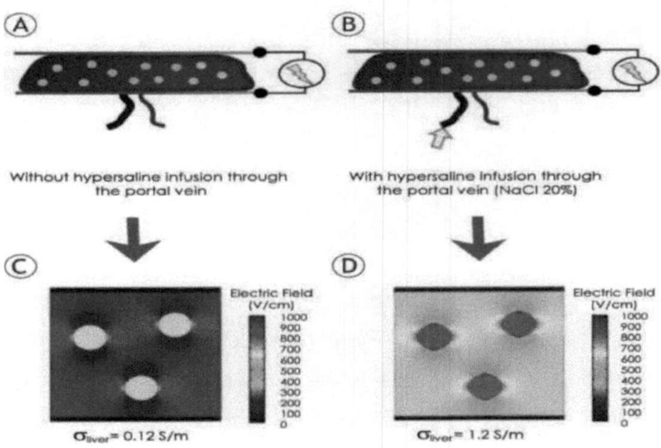

Fig. 4: Schematic diagram of the mechanism of action of electroporation.

Appelbaum et al show that applying 4 electrodes in parallel achieves a greater ablation volume with better oncologic results, since in this way the protocol can be personalized for each patient (Fig. 5). In clinical practice, an electric field that varies between 1,000-1,500V/cm is usually used.

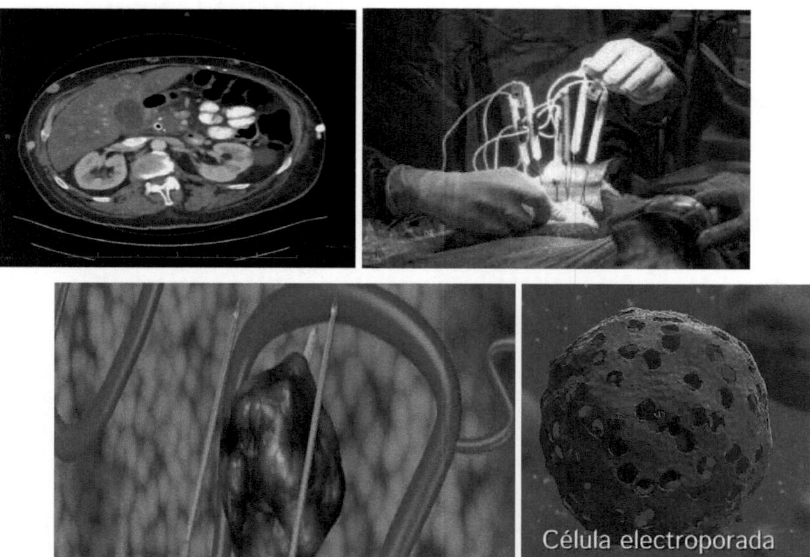

Fig. 5: *CT image, technique and result of electroporation.*

There is no established limit on the size of the lesion to be treated, but better oncologic results are described in lesions smaller than 3cm and when the tumor volume exceeds 5cm^3 a higher risk of local recurrence is described. The efficacy and feasibility of IRE in perivascular ablation is one of its great advantages, having reported ablations with an average proximity to the great blood vessels of <0.5cm, without observing thrombosis or stenosis in the follow-up (10,11).

ADVANTAGES AND DISADVANTAGES OF THE IRE TECHNIQUE.

It is a growing technology that is likely to have a splendid future in the coming years due to the development, among other measures, of new and more powerful generators that are currently in the design phase. (12) The possible benefits and drawbacks of irreversible electroporation are shown in Table 1 (13).

Table 1: Advantages and disadvantages of the IRE technique

Ventajas:

- Naturaleza no térmica, lo que permite la ablación en estructuras vitales o cerca de ellas.
- Eliminación de los efectos del calor y el frío.
- Obtención de imágenes (TC o ecografía) en tiempo real durante el procedimiento de la zona sometida a ablación.
- Zona de ablación identificable debido a la precisión en la localización tumoral.
- Posibilidad de realizar la ablación de múltiples lesiones o múltiples ablaciones de una sola lesión en una única sesión.
- Dolor escaso o inexistente después del tratamiento debido a la escasa inflamación que se produce en el procedimiento.
- Resolución rápida de la lesión, debido a la cicatrización que se produce, con células que delimitan y eliminan las dañadas.
- Menos eventos adversos con preservación del tejido.

Desventajas:

- Generación de pulsos eléctricos con posibilidad de estimular la contracción muscular o inducir arritmias cardíacas.
- Riesgos asociados a la anestesia general y a la parálisis muscular.
- Recursos, tiempo y costes requeridos para la anestesia general.
- Elevada velocidad del procedimiento, lo que impide posibles ajustes del tratamiento durante el procedimiento; el pulso es generado y aplicado instantáneamente.
- Necesidad de elevada precisión en la colocación de los electrodos.
- Riesgo de sangrado, a causa de la no posibilidad de coagulación en la zona próxima a la inserción de los electrodos.

Tumor volume and underlying tumor disease were significantly associated with local recurrence according to Niessen, however IRE ablation of perivascular liver lesions is not associated with early local recurrence after IRE (14).

INDICATIONS AND CONTRAINDICATIONS

Irreversible electroporation is currently only indicated for tumors that are not suitable for surgical resection and thermal ablation. Frequently, this applies to centrally located liver tumors. Irreversible electroporation can be administered more than once, and can be used to treat both residual disease and new lesions; being a non-thermal technique, it can be performed open or percutaneously (with the aid of ultrasound or CT). The indications and contraindications are described in Table 2.

Table 2: indications and contraindications for IRE

Indications IRE	Contraindications IRE
Tumors unsuitable for surgical resection or thermal ablation	Patients with cardiac implants, pacemakers or defibrillators
Centrally located tumors	In injuries near implanted electronic devices
Treatment of residual disease and new lesions	In injuries near implanted devices with metallic parts
Treatment of lesions 3-4 cm in length	Patients with epilepsy, cardiac arrhythmias or recent myocardial infarction or patients with coagulopathy.
It can be performed percutaneously or open (CT echo).	**Impossibility of general anesthesia or neuromuscular block.**

COMPLICATIONS.

IRE ablation has a number of side effects and complications as shown in Table 3 (15). and complications, which are shown in Table 3.

Table 3: Complications of IRE ablation

General complications	Local complications
Cardiac arrhythmias	Pneumothorax (3.9%)
Transient increase in AT	Pleural effusion
Hyperkalemia	Hematoma at site of action (12%)
Metabolic acidosis	Muscle hyperstimulation pain

BIBLIOGRAPHY .

1. Miller L, Leor J, Rubinsky B. Cancer cells ablation with irreversible electroporation. Technol Cancer Res Treat. 2005; 4: 699-705.

2. Scheffer H.J, Nielsen K, et al. Irreversible electroporation for nonthermal tumor ablation in the clinical setting: A systematic review of safety and efficacy. J Vasc Interv Radiol.2014; 25: 997-1011.

3. Deodhar A, Dickfeld T, Single GW, Hamilton WC, Jr, Thornton RH, Sofocleous CT, et al.Irreversible electroporation near the heart: Ventricular arrhythmias can be prevented with ECG synchronization. *AJR Am J Roentgenol.* 2011; 196:330-5.

4. Jiang C, Davalos RV, Bischof JC. A review of basic to clinical studies of irreversible electroporation therapy. *IEEE Trans BioMed Eng.* 2015; 62:4-20.

5. Knavel E.M., Brace C.L.: Tumor ablation: common modalities and general practices. Tech Vasc Interv Radiol 2013; 16: 192-200.

6. Scheffer H.J., Nielsen K., van Tilborga J.M., et al.: Ablation of colorectal liver metastases by irreversible electroporation: results of the COLDFIRE-I ablate-and-resect study. Eur Radiol 2014; 24: 2467-2475.

7. Niessen C, Beyer L.P., Pregler B, Dollinger M, Trabold B, Schlitt H.J, et al. Percutaneous ablation of hepatic tumors using irreversible electroporation: A prospective safety and midterm efficacy study in 34 patients. J Vasc Interv Radiol. 2016; 27: 480-486

8. Lee E.W., Thai S., Kee S.T.: Irreversible electroporation: A novel image- guided cancer therapy. Gut Liver 2010; 4: S99-S104.

9. Golberg A, Bruinsma B.G, Uygun B.E, Yarmush M.L. Tissue heterogeneity in structure and conductivity contribute to cell survival during irreversible electroporation ablation by "electric field sinks". Sci Rep, 2015; 5: 8485.

10. Chen X, Ren Z, Zhu T, Zhang X, Peng Z, Xie H, et al. Electric ablation with irreversible electroporation (IRE) in vital hepatic structures and follow-up investigation. Sci Rep, 2015; 9: 16233.

11. M. Dollinger, R. Müller-Wille, F. Zeman, M. Haimerl, C. Niessen, L.P. Beyer, et al. Irreversible electroporation of malignant hepatic tumors - Alterations in venous structures at subacute follow-up and evolution at mid-term follow-up. PLoS One. 2015; 10: e0135773

12. Sánchez-Velázquez P, Q. Castellví, A. Villanueva, R. Quesada, C. Pañella, M.

Cáceres, et al. Irreversible electroporation of the liver: Is there a safe limit to the ablation volume? Sci Rep. 2016; 6: 23781.

13. ECRI Institute. Irreversible Electroporation (NanoKnife System) for Treating Malignant Solid Primary Tumors and Metastases to the Liver. Emerging Technology Evidence Report. Plymouth Meeting, PA: Ecri Institute; 2013.

14. Niessen C, Igl J, Pregler B, Beyer L, Noeva E, Dollinger M, et al. Factors associated with short-term local recurrence of liver cancer after percutaneous ablation using irreversible electroporation: A prospective single-center study. J Vasc Interv Radiol.2015; 26: 694-702.

15. Kambakamba P, Bonvini J.M, M. Glenck M, Castrezana Lopez L, Pfammatter T, Clavien P.A, et al. Intraoperative adverse events during irreversible electroporation - A call for caution. Am J Surg. 2016; 212: 715-721.

Chapter 5: Cryotherapy Ablation

INTRODUCTION

Cryotherapy is one of the ablation methods used to destroy liver metastases (1). This method requires the placement of a special probe near the cancer site. The probe is used to deliver extreme cold to the site, which is produced by liquid nitrogen or argon gas. The placement of the probe may be guided with ultrasound or computed tomography. The flash freezing process destroys the cancer cells and shrinks the cancer. However, it is unclear whether this treatment prolongs life or increases the quality of life of patients. It was first described in 1963; however, it has been used more frequently after the development of ultrasound that allows better monitoring of the effect of therapy.

It was one of the first ablative methods used in hepatocarcinoma and also in hepatic metastases, but the larger diameter of the probes (which usually require laparotomy for their application) and the number and severity of the complications that have occurred with this method have made it fall into disuse.

TECHNIQUE AND MECHANISM OF ACTION

It consists of the destruction of tissue by the application of very low temperatures (-20º to -50º) leading to freezing / necrosis.

The phenomenon of freezing causes ice crystals to form (intra- and extracellular).

- Intracellular crystals lead to cell death by injury to cell membranes, structures within the cell, or both when cooling occurs rapidly or at very low temperatures.

- Extracellular crystals occur when cooling is slow and causes cell death due to changes in osmotic gradients.

— Another mechanism of freezing is the production of local ischemia by thrombosis of small vessels.

The technique involves the use of probes through which liquid nitrogen is circulated at temperatures below -196ºC. Cryotherapy is based on the formation of intracellular crystals at temperatures of -35ºC, which induce cell necrosis and its effect is intensified by freeze-thaw cycles (Fig. 1).

The mechanism of action: During freezing, ice crystals (intracellular and extracellular) are formed. Intracellular crystals lead to cell death by injury to cell membranes, structures within the cell, or both. The formation of these cellular crystals only occurs when cooling occurs rapidly or at very low temperatures; the phenomenon is adjacent to the needle. The formation of extracellular ices occurs when cooling is slow causing cell death by changes in osmotic gradients. Initially, cooling dehydrates the cells, and when the tissue is heated, it edematizes and explodes. Finally, freezing produces local ischemia by thrombosis of small vessels. The temperature that must be reached to achieve necrosis is -20º to -50º. The "radiator" phenomenon prevents large caliber vessels from freezing; neither do the vessel walls suffer damage (2).

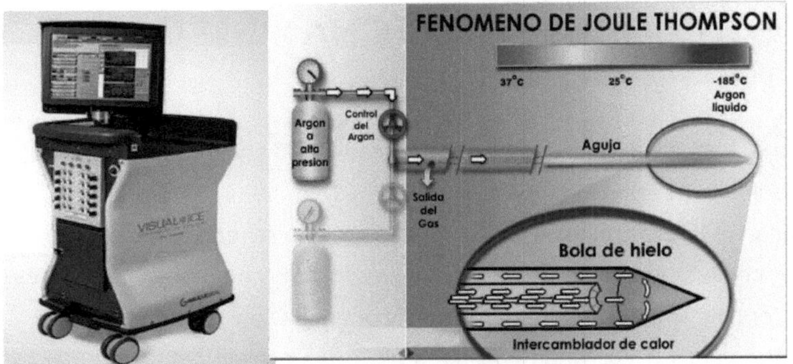

Fig. 1: Cryoablation technique

There is no consensus on the number of lesions to be treated, although 4 to 6 lesions are recommended. The size of the lesions should not exceed 5 cm. The percutaneous approach is the most favorable with the best results (Fig. 2).

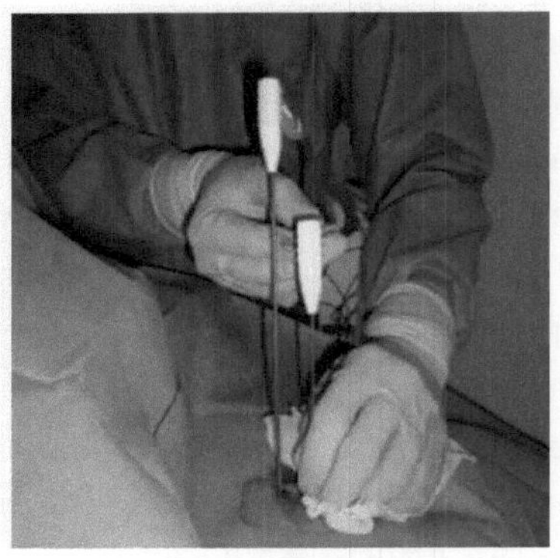

Fig. 2: Percutaneous placement of cryoprobes

INDICATIONS AND CONTRAINDICATIONS

1. *Indications:*

- Cryotherapy may be indicated when metastases are limited to the liver.

- In patients with apparently resectable lesions but extensive cardiac or pulmonary comorbidities and limited hepatic reserve.

- In anatomically unresectable lesions.

- Bilobular lesions and lesions in close proximity to major blood vessels or bile ducts.

2. *Contraindications:*

- No absolute contraindications for cryotherapy have been defined.

- We consider it appropriate for a patient with anatomically resectable lesions with little or no comorbidity. anatomically resectable lesions with little or no comorbidity.

- It is also not indicated for patients with extrahepatic metastases.

- It is relatively contraindicated when the number of lesions is greater than 6.

COMPLICATIONS

Complications of cryotherapy can be mild or severe (3,4) as shown in Table 1.

Table 1: Complications of liver cryoablation

Mild complications	Serious complications
Pleural effusion	Thrombocytopenia
Hemorrhage	Disseminated intravascular coagulation
Myoglobinuria with . acute tubular necrosis.	Renal failure
Hemoperitoneum	Craking (hepatic freezing burst)
Biliperitoneum	Cryoshock leading to MOF and coagulopathy.
Hepatic abscesses	
Fistulas and biliary strictures	
Elevated liver function tests liver function tests	

RESPONSE EVALUATION

It is recommended to follow up patients every 3-6 months with CT, the lesions are best evaluated during the arterial phase, in which a hypodense image is observed after cryotherapy (5) and later in the follow-up a hyperdense halo in the cryoablation area (6), with ultrasound it is appreciated with acoustic shadow.

The morbidity of cryotherapy is between 15-20%, with a mortality of up to 4%. Survival is up to 27-30 months (7).

ADVANTAGES AND DISADVANTAGES OF CRYOABLATION

The advantages and disadvantages (8, 9.10) are described in Table 2.

Table 2: Advantages and disadvantages of cryoablation

ADVANTAGES OF CRYOABLATION	CRYOABLATION DISADVANTAGES
Shorter recovery time	High cost
It has an anesthetic effect. Usually requires little anesthesia	Limited indications
Respects medium and large caliber vascular structures.	Longer time required to reach freezing during the technique
Multiple multiple lesions simultaneously percutaneously	
Maintains integrated protein structure, which favors immune response (IL-2, IL-6...)	
Correct visualization of the area to be treated (ice ball).	

BIBLIOGRAPHY .

1. Goldeberg et al. Image-guided tumor ablation.Standardization of terminology and reporting criteria. J.Vasc. Radiolol 2009; 20: 377-390.

2. Gage A, Baus JM, Baus JG. Experimental cryosurgery Investigations in vivo. Cryobiology 2009; 59: 229

3. Meyers M, Sasson A, Sigurdson E. Locoregional strategies for colorectal hepatic metastases. Clinical colorectal cancer. 2003; 3:34-44

4. Seifert JK, France MP, Zhao J, et al. Large volume hepatic freezing association with significant release of the cytokines interleukin-6 and tumor necrosis factor a in a rat model. World J. Surg 2002; 26: 1333- 1341.

5. Fortner JG, Blumgart Lh. A historic perpective of liver surgery for tumors at the end of the millennium. J. Am. Coll Surg 2001; 193: 210-222.

6. Rewcastle JC, Sandison GA, Saliken JC, Donnelly BJ et al. Considerationduring clinical operation of two commercial avaible cryomachines. J. Surg Oncol 1997; 71: 106-111.

7. Terence C et al. Outcomes of a single-centre experience of hepatic resection and cryoablation of sarcoma liver metastases. J. Clin Oncol 2011; 34: 317-320.

8. Cooper SM, Dawber RP. The history of cryosurgery. J. R. Soc Med 2001; 94: 196-201.

9. Theodorescu D. Cancer cryotherapy: evolution and biology. Rev Uro 2004; 6 Suppl 4: S9-S19.

10. Sung GT, Gill IS, Hsu TH et al. Effect of intentional cryo-injury to the renal collecting system. J. Urol 2003; 170: 619-622.

Chapter 6: High-frequency ultrasound ablation intensity (HIFU)

INTRODUCTION

The acronym HIFU (High-Intensity Focused Ultrasound) is a type of technique or procedure based on the application of sonic waves focused and directed specifically to a target area or objective, in order to cause the death or necrosis of certain cells of living organisms (1).

High intensity focused ultrasound (HIFU) ablation is a non-invasive extracorporeal treatment method using focused ultrasound beams that is capable of producing complete coagulative necrosis of target lesions through intact skin without surgical exposure or instrumentation (2).

The application of High Intensity Focused Ultrasound (HIFU) is a

known transcutaneous ablation system similar to extracorporeal lithotripsy, these mechanical waves aim to achieve localized temperature elevation and necrosis by complete coagulation of the tumor in as few sessions as possible. When applied to the liver, the

Transcutaneous HIFU technology has had to overcome considerable challenges such as respiratory movements, distortion of the mechanical wave beam by the ribs and long application times. Nevertheless, recent technological advances make it an attractive focal ablation system with considerable experience (3).

This procedure is non-invasive, since it does not require surgery or chemical elements, and has the advantage that it does not damage the tissues between the ultrasound emission site and the target area.

TECHNIQUE AND MECHANISM OF ACTION

The general principle of HIFU involves ultrasound beams with short wavelengths and megahertz frequencies focused on small volume target areas, resulting in immediate coagulative necrosis of the target tissue. The temperature of the area outside the target lesion is below 55 °C, which is not sufficient to induce thermal injury. In addition, the location and extent of treatment can be accurately monitored with real-time ultrasound. Theoretically, HIFU damage is assumed to be confined within the intended ablation area without damaging overlying and adjacent structures, even within the beam path (4).

HIFU ablation can be performed by a single exposure, resulting in ellipsoidal-shaped focal ablation zones or rectangular-shaped row ablation zones formed by combining several ellipsoidal ablation zones. Since the ablation zone caused by a single ultrasonic exposure is very small, multi-row ablation zones are mainly applied until the entire target area is covered (Fig. 1).

The operation of this technique bases its effectiveness on the fact that the sound waves, concentrated in an area when applied in a beam, end up generating heat energy that produces hyperthermia in the target areas. It also generates a mechanical force, in the form of vibration, which allows the tissues to be compressed or decompressed.

Fig. 1: Image-guided ablative HIFU treatment

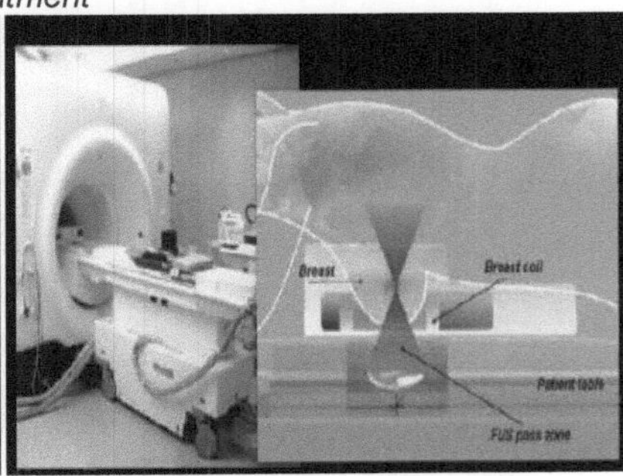

This technique reduces the volume of the lesion and sometimes eliminates it completely, so it is necessary to combine it with other treatments (chemotherapy/immunotherapy) with which a synergistic effect is generated, improving the survival of our patients (5).

By means of a device that focuses a high-intensity ultrasound beam, the maximum possible tumor volume is thermally destroyed. Throughout the procedure, the treated area is visualized by ultrasound or CT scan, thus guaranteeing the effectiveness of the treatment in real time and reducing the possibility of adverse effects or complications to a minimum.

The HIFU technique applies temperature between 55 to 80ºC for a few seconds in a focused manner to cause cell death. The monitoring of temperature changes provides three-dimensional images in the target tissue, which minimizes the risk of damage to adjacent tissues and the possibility of adverse effects.

This technique reduces the volume of the nodular lesion and sometimes eliminates them completely as it is non-invasive, does not require surgery or chemical elements.

Currently, HIFU treatment is attempted for solid tumors in different areas of the body. Focused ultrasound can cause the focal temperature to rise to over 90°C rapidly and create cavitation effects, resulting in coagulation necrosis of the target tissue and damaged blood vessels within the lesions (6).

The benefits - advantages of the treatment are (7):

- Death and destruction of tumor cells by thermal injury.
- Reduce the volume of the lesion and sometimes eliminate it altogether
- The most relevant advantage of non-invasive surgery is that it makes it possible to eliminate a large part or all of the tumor lesion without any aggression to adjacent tissues. This allows a very short hospital stay with a rapid recovery of the patient.
- Stimulates antitumor immune response
- It also reduces the risk of infection or introduction of toxic elements.
- May improve quality of life, reduce or eliminate tumor-related pain

INDICATIONS AND CONTRAINDICATIONS

Recently, sustained response percentages of more than 30% in unresectable primary and secondary tumors have been reported in several Eastern series with the new ultrasound beam focusing systems. Nevertheless, there are as yet no studies of proven quality to adequately evaluate this technology, so its application in the West is irregular. The indications and contraindications of HIFU are (8):

1. *Indications:*

- Adjuvant treatment with other therapies
- Treatment of relapses or even as palliative treatment.
- When there are transfusion rejections.
- When surgical resection is not possible.

2. *Contraindications:*

- People who have autoimmune problems or an altered or weakened immune system.
- People with open wounds
- Coagulation disorders
- Patients with febrile processes or in a hyperthermic state.
- Patients with esthetic implants (heat can cause them to be reabsorbed or generate severe burns), metallic implants at least in the area to be treated or nearby.
- Patients with pacemakers (given the risk that sonic waves may affect the implant) and metallic implants.
- Renal insufficiency
- Severe diabetes or metabolic diseases.
- Neither in areas such as lung, stomach or intestine since the gas they contain limits their effect.

COMPLICATIONS

Complications arising from HIFU treatment for liver metastases can be divided into two general categories (9) and can be of 3 types as described in Table 1.

- Those related to thermal injury adjacent to the organ or ultrasound beam pathway, due to shallower areas of the target lesion or unwanted deep penetration over target areas.
- Related to artificial pleural effusion.

Table 1: Complications of high-intensity ultrasound ablation

C. MAJORS	C. MINORS	C. LATE
• biliary obstruction • symptomatic pleural effusion • pneumothorax • wall-to-tumor fistula • intestinal perforation • third degree burns	• skin thickening, erythema, erythema • muscle pain/abdominal pain • nausea • first and second degree burns • asymptomatic pleural effusion • dyspnea and fever • tingling sensation	• diaphragmatic rupture • rib fracture

BIBLIOGRAPHY .

1. Zhou YF. High-intensity focused ultrasound in clinical tumor ablation. World J Clin Oncol. 2011; 2: 8-27.

2. Kennedy JE, Ter Haar GR, Cranston D. High-intensity focused ultrasound: surgery of the future? J Radiol. 20 03; 76: 590-599.

3. Zhang L, Wang ZB. Ablation of tumors with high-intensity focused ultrasound: review of ten years of clinical experience. Front Med China 2010; 4:294-302.

4. Haar GT, Coussios C. High-intensity focused ultrasound: physical principles and devices Int J Hyperthermia. 2007; 23: 89-104

5. Huber, P., Debus, J. & Jenne, J. (1996). Therapeutic ultrasound in tumor therapy: principles, applications and new developments. Radiology, 36: 64-71.

6. Chapelon, J.Y., Margonari, J. & Vernier, F. (1992). In-vivo effects of high intensity ultrasound on prostatic adenocarcinoma Dunning R3327. Cancer Res., 52: 6353-6357.

7. Vidal-Jove, J; Perich, E; Jaen, A, Alvarez del Castillo, M: Hyperthermic ablation by UltraSound Guided High Intensity Focused Ultrasound.

(USgHIFU) plus Systemic Chemotherapy (SC) for locally advanced pancreatic cancer: the secret of longer survival. Journal of Therapeutic Ultrasound 2014, 2 (suppl 1) A6.

8. Dubinsky TJ, Cuevas C, Dighe MK, Kolokythas O, Hwang JH High-intensity focused ultrasound: current potential and oncological applications. J Roentgenol. 2008; 190:191-199

9. Li JJ, Xu GL, Gu MF, et al. (2007) Complications of high-intensity focused ultrasound in patients with recurrent and metastatic abdominal tumors.J. Gastroenterol 2007; 13: 2747-2751.

Chapter 7: Nanothermal Ablation

INTRODUCTION

Oncological Hyperthermia is a technique based on the treatment of tumors with high temperatures. The use of heat therapies in the treatment of cancer (hyperthermia) is not something new, but its benefits were well known since ancient Greece as Hippocrates and frequently used by Arab medicine.

Oncothermia was founded in 1988 by Szasz (1); it allows the focused action of this technique and manages to significantly improve the life prospects of the cancer patient. Its main function is to generate a fractal electro alternating field (without emitting any radiation) thanks to two electrodes: a fixed one located on the surface of the bed on which the patient lies, and a mobile one through an articulated arm that is focused on the primary tumor or metastases. This electro field manages to permanently increase the temperature of the local tumor area to be treated and supplies constant energy; thus generating a temperature mismatch between the extra and intracellular electrolytes until thermal equilibrium is reached at the end of the treatment.

Oncothermia, also known as "electromodulated hyperthermia", "mEHT or Nanothermia in Spain, is a method of electro modulated hyperthermia, non-invasive, adjuvant in the treatment against cancer. It is a treatment that intervenes selectively and directly on cancer cells, inhibiting their activity, stimulating the patient's immune response, causing their destruction, and even effectively helping to reduce the patient's pain.

The Nanothermia system is effective in the treatment of solid and inoperable tumors, since it offers the possibility of reaching the tumor in a deep and localized manner. Each tumor will receive a personalized treatment and the appropriate variables will be applied according to its particular characteristics (2).

TECHNIQUE AND MECHANISM OF ACTION

It is based on the application of heat from outside the patient thanks to a state-of-the-art medical device. The Oncothermia device generates a dielectric field of energy capable of causing a significant increase in temperature (up to 45ºC) deep inside the tumor. This energy is absorbed by the extracellular fluid of the malignant cells and damages them considerably, causing their destruction (3).

The main reason for these good results lies in the fact that tumor tissues have higher conductivity than healthy tissues (due to metabolic differences at the cellular level); and that cancer cells are more sensitive to heat therapies. Under this premise, Oncothermia acts in a focused manner on tumor tissue and is absorbed without affecting healthy cells.

It is an effective treatment in monotherapy, but it is advisable to use it in combination with other systemic treatments such as radiotherapy and/or chemotherapy due to its synergistic action, since this increase in temperature increases the oxygenation of the cancer cells (whose metabolism is normally hypoxic or low in oxygen), making them more sensitive to chemo- and/or radiotherapy treatments, thus enhancing their effects and considerably reducing the harmful sequelae derived from them(4).

This system generates a fractal electro alternating field by means of two poles, a fixed one located in the area of the bed on which the patient lies and a movable one through an arm that is placed in the position of the primary tumor or metastasis. In this area an absolute increase in temperature develops, derived from the energy applied. There is no radiation (Fig. 1). Thus, a permanent temperature increase is achieved in the tumor area. This temperature difference acts on the cell membrane and leads to destabilizing thermal stress on the tumor cell membrane, leading to apoptosis (programmed cell death).

This technique, unlike other treatments based on hyperthermia (4), does not heat the tumor, the metastasis and the surrounding tissue, with the side effects that this heating would cause in the patient, but applies heat to the tumor using electromagnetic energy; it achieves a permanent increase in temperature in the tumor area, without radiation and, thanks to the constant energy supply, a temperature gradient is generated between the extra and intracellular electrolytes until thermal equilibrium is reached at the end of the therapy. This temperature difference (although in absolute numbers is low),   acts on the cell membrane and this leads to a destabilizing thermal stress on the tumor cell membrane (6), leading to apoptosis or programmed cell death. Each tumor, depending on its histological type and other variables, will receive its own time and dose, will have a different treatment and variables adapted to its characteristics will be applied. There is a control mechanism on the absorbed energy dose of the tumor. It does not therefore use temperature as a parameter, but the "specific absorbed energy" (Gy).

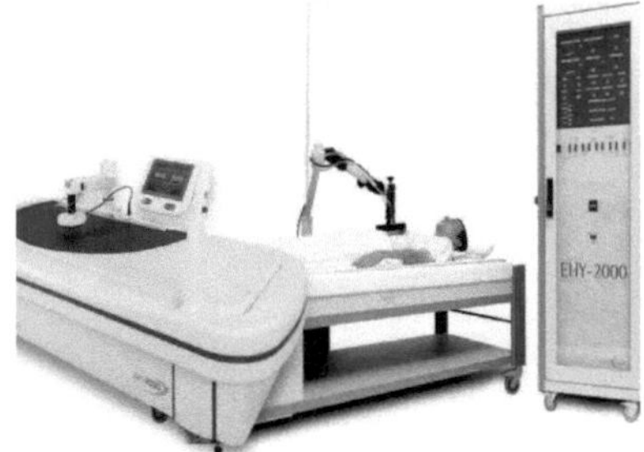

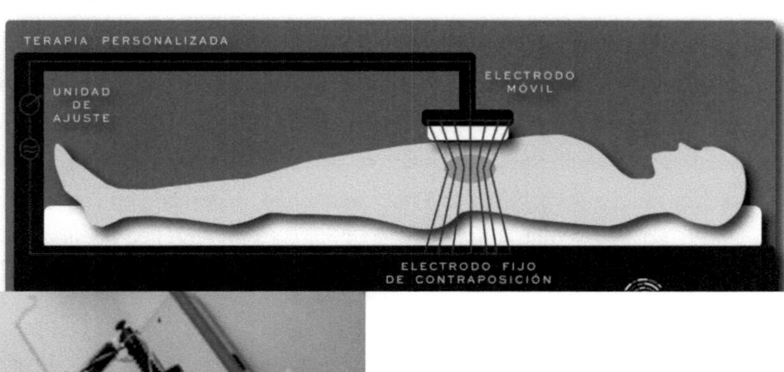

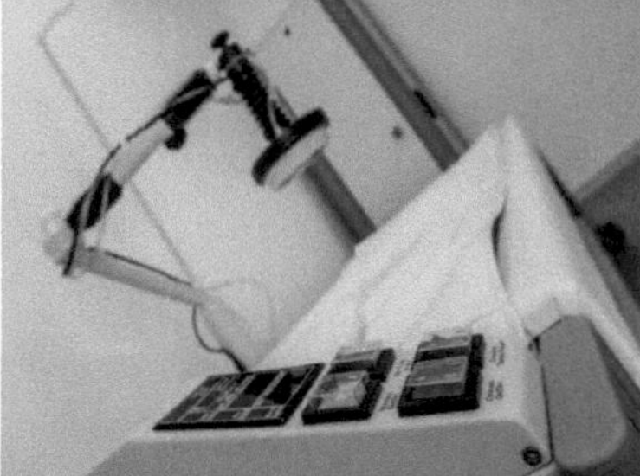

Fig. 1: Medical nanothermia system (top). Schematic illustration of the EHY-2000 Oncothermia device and how it acts on the patient (below).

Oncothermia is a therapy that acts mainly at two levels:

A. By means of a permanent selective temperature increase in the extracellular fluid of the tumor tissue. A modulated electric field is generated which transfers energy in a constant manner thanks to the principle of capacitive coupling (like a capacitor) of radiofrequency waves of 13.56 MHz. This field allows energy absorption in the extracellular fluid (ECM). Due to the constant energy input, a temperature gradient is generated between the extra- and intracellular electrolytes until a thermal equilibrium is reached at the end of the treatment.(6)

This temperature difference acts through the cell membrane of the cancer cell (from the outside to the inside) causing thermal stress that destabilizes and weakens it, boosting the patient's immune system and causing apoptosis or programmed cell death.

B. By increasing the oxygenation of malignant cells. One characteristic that makes cancer cells resistant to cancer treatments is their degree of hypoxia (oxygen deficiency), which is much higher than that of healthy cells. Oncothermia achieves this increase in the oxygenation of cancer cells, thus making them more vulnerable to chemo and/or radiotherapy treatments. (7)

International Oncothermia protocols usually recommend 2 to 3 sessions per week, leaving one day of rest between them, up to a total of 12 sessions, coinciding with chemotherapy and radiotherapy treatment or in monotherapy. The time of each session usually ranges from 1 hour to 1 hour and a half, depending on the patient's pathology. It is advisable that at least 6 weeks elapse since the last application of the Oncothermia treatment, to check the result in medical imaging tests; in addition to progressively controlling the tumor markers until the test is performed.

The effects of Oncothermia on the patient are:

• Increased destruction of cancer cells, by affecting the tumor cell membrane, improving the induction of apoptosis. Thus inhibiting tumor growth, producing alterations in the neoplastic cell cycle.

• An increase in vascular temperature (8), which implies greater vasodilatation and increased tissue oxygenation, reducing hypoxia and pH acidity. These conditions increase the sensitivity of the neoplastic tissue to the action of chemotherapy and radiotherapy treatments combined with oncothermia.

• Reduction of pain in the patient.

Effects at the cellular level.

1. The tumor is subjected to a modulated electric field specific to its characteristics. The electric field is applied specifically on tumor cells without affecting healthy tissue (9).

2. The electric field affects the biochemical processes of the tumor, affecting the plasma membranes of the tumor cells, modifying their transmembrane potential, which generates intra- and extracellular reactions: the intracellular concentration of sodium increases and there is an outflow of potassium.

3. As a consequence of this change in the membranes and their potential, the formation of cellular connections is promoted (molecules called catenins and E-cadherins with activation of p53) which re-establish order and provoke cell collapse or apoptosis. The treatment lasts between an hour and an hour and a half depending on the type of tumor (10,11).

4. All this also generates effects not only on the tumor but also indirectly by facilitating the immune system's contact with tumor antigenic structures. The immune system becomes more competent to fight cancer (12).

Benefits.

- Non-invasive treatment without side effects. It is selective, and if applied together with other classical techniques (chemotherapy/radiotherapy), it reduces the harmful consequences of the latter and improves the efficacy of the antitumor treatment.

- It increases the therapeutic effects of chemotherapy and radiotherapy; it can be applied with them, before or after surgery. By eliminating malignant cells (through apoptosis), it offers the immune system the possibility of identifying them and generating more effective responses both in a primary tumor and in metastases.

- Reduces the size of tumors at a very advanced stage of development with conventional means

- Improves immune response by eliminating malignant cells by apoptosis

- It sometimes facilitates surgery and can be applied after surgery.

- Pain reduction. It is a painless treatment. In addition, it leads to increased immunogenicity and effectively reduces the patient's pain.

- Increases vascular temperature and allows conventional therapies such as chemotherapy and radiotherapy to be more effective in poorly vascularized areas where there is a lack of oxygen, since it vascularizes resistant areas and eliminates hypoxic areas. (8)

- Increased capacity to destroy tumor cells by inducing apoptosis, a molecular mechanism that occurs inside the cells by which our organism eliminates unnecessary, infected and potentially cancerous cells. This is a form of programmed cell death as opposed to necrosis, where cell death is caused by damage or injury (13).

- The Nanothermia system adapts its modulation to the patient's type of cancer. Each tumor will receive a different treatment and variables adapted to its characteristics will be applied.

Contraindications.

Contraindications to this treatment must be evaluated on a case-by-case basis and are described in Table 1.

Table 1: Contraindications of nanothermia - Oncothermia

CONTRAINDICATIONS
Transplanted or immunosuppressed patients
Pregnant women
Unconscious patients, without physical sensibility
Patients with pacemakers or electronic implants
Patients with metallic prosthesis or metallic reservoir (Port-a-Cath) close to the application area.
In patients with open wounds, in the area to be applied.
Patients with epilepsy
People who have had silicone implants.
If there is pleural effusion, ascites or other types of fluid accumulation.

BIBLIOGRAPHY .

1. Szasz, "Physical background and technical realization of hyperthermia," in Locoregional Radiofrequency-Perfusional- and Wholebody- Hyperthermia in Cancer Treatment: New Clinical Aspects, G. F. Baronzio and E. D. Hager, Springer Science and Eurekah.com. 2006; Eds., pp. 27-59.

2. Szasz, N. *Oncothermia-Principles and Practices*, Springer, 2010.

3. Singh, "Hyperthermia-a new dimension in cancer treatment," Indian Journal of Biochemistry and Biophysics, 1990; vol. 27 (4): 195-201.

4. M. Israël and L. Schwartz, "The metabolic advantage of tumor cells," Molecular Cancer, vol. 10, article 70, 2011.

5. H. R. Moyer and K. A. Delman, "The role of hyperthermia in optimizing tumor response to regional therapy," *International Journal of Hyperthermia*, vol. 24, no. 3, pp. 251-261, 2008.

6. G. Vincze, N. Szasz, and A. Szasz, "On the thermal noise limit of cellular membranes," *Bioelectromagnetics*, vol. 26, no. 1, pp. 28-35, 2005.

7. M. R. Horsman and J. Overgaard, "Can mild hyperthermia improve tumour oxygenation?" *International Journal of Hyperthermia. 1997;* vol. 13 (2): 141-147.

8. P. W. Vaupel and D. K. Kelleher, "Pathophysiological and vascular characteristics of tumours and their importance for hyperthermia: heterogeneity is the key issue," *International Journal of Hyperthermia*, 2010; vol. 26 (3): 211-223.

9. V. D. Ferrari, S. De Ponti, F. Valcamonico et al, "Deep electro- hyperthermia (EHY) with or without thermo-active agents in patients with advanced hepatic cell carcinoma: phase II study," *Journal of Clinical Oncology* 2007; vol. 25: 18S.

10. Gadaleta-Caldarola G, Infusino S, Galise I, et al. (2014) Sorafenib and locoregional deep electrohyperthermia in advanced hepatocellular carcinoma. A phase II study. Oncol Lett, 2014; 8(4):1783-1787.

11. R. M. Bremnes, R. Veve, F. R. Hirsch, and W. A. Franklin, "The E- cadherin cell-cell adhesion complex and lung cancer invasion, metastasis, and prognosis," *Lung Cancer*, 2009; vol. 36(2): 115-124.

12. Andocs G., Szasz O., Szasz A. Oncothermia Cancer Treatment: From the Laboratory to the Clinic. electromagnetic Biol. Medicine. 2009; 28:148-165.

13. Andocs G., Renner H., Balogh L., Fonyad L., Jakab C., Szasz A. Strong synergy of heat and modulated electromagnetic field in tumor cell destruction. Strahlenther. Onkol. 2009; 185:120-126.

INTRODUCTION

Direct thermal destruction of liver tumors by laser energy is known by several acronyms: laser thermal ablation (LTA), interstitial laser thermotherapy (ILT), interstitial laser photocoagulation (ILP) and LITT.

Laser interstitial thermal therapy (LITT) is a minimally invasive technique for the treatment of tumors. The concept has been around since the late 1970s, and in 1983 Bown et al (1) used a neodymium-doped yttrium aluminum garnet (Nd:YAG) laser initially for treatment of brain tumors. Subsequently the indications have been expanded and it is now also used in the treatment of liver metastases. Laser interstitial thermal therapy (LITT) offers a minimally invasive surgical option for treating different types of tumors.

The main biological effect of interstitial laser irradiation is thermal damage. (2) LITT results in heating of the treated tissue and causes enzyme induction, protein denaturation, membrane lipid fusion, vessel sclerosis and coagulation necrosis. Heat-induced tissue necrosis results from a temperature rise above 43°C, and the time to cell death is exponentially dependent on temperature. Rapid increases in temperature can cause tissue charring, which subsequently changes the optical properties of the tissue and limits laser penetration. (3) Overheating can also result in tissue vaporization.

TECHNIQUE AND MECHANISM OF ACTION

Laser-induced interstitial thermotherapy (LITT), or interstitial laser photocoagulation, is similar to hyperthermia t r e a t m e n t , which uses heat to shrink tumors by damaging or destroying cancer cells. During LITT, an optical fiber is inserted into a tumor. Laser light at the tip of the fiber raises the temperature of the tumor cells and damages or destroys them. Laser-induced thermotherapy (LITT) is minimally invasive; it is a gentle therapy method for the local treatment of malignant liver tumors.

The technique is performed under local anesthesia using CT or ultrasound (percutaneously), the laser applicator is pushed through a small hole in the abdominal skin directly into the liver tissue. A laser probe is then placed under the skin into the liver tissue. This laser light with a wavelength of 1064 nanometers is transmitted through a glass fiber to the tumor cells, which heats and destroys them. Magnetic resonance imaging (MRI) closely monitors and controls the treatment and the progress of the therapy can be evaluated with the help of special temperature-sensitive images. The duration of the procedure is approximately half an hour. During this time, the patient is awake.

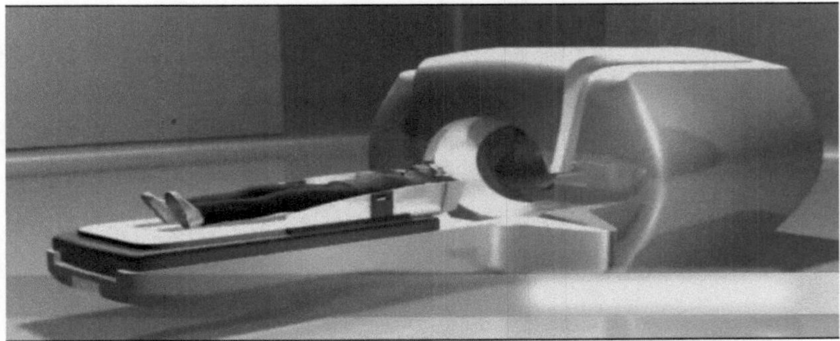

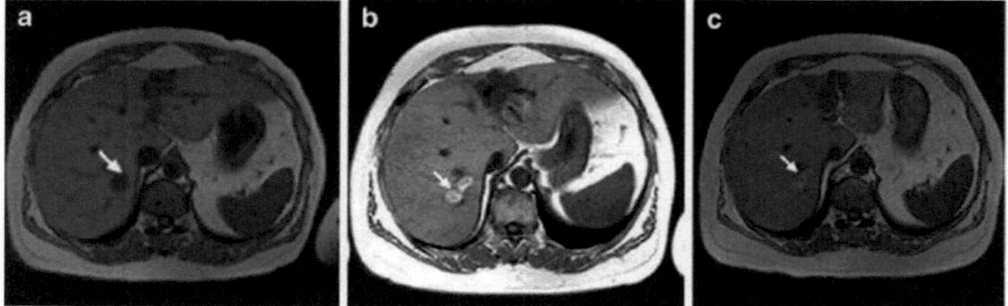

Fig. 1: Upper: *Interstitial laser system with MRI.* ***Bottom*** *T1-weighted MR images in patient with liver metastases treated with laser-induced thermotherapy (a) before, (b) 24 h after surgery and (c) 6 months after laser-induced thermotherapy.*

LITT uses laser energy delivered to a target through a fiber-optic catheter, which damages intracellular proteins and DNA, and subsequent cell death. LITT provides a clear demarcation between ablation and unharmed tissue, which, when combined with the ability to monitor and control the ablation through MR thermal imaging, results in a high level of precision and control. Coagulation is achieved using a neodymium yttrium aluminum garnet laser light. The multi-applicator technique involves treating a lesion with multiple (up to 5) laser applicators simultaneously (4).

This is a tissue destruction procedure, using the heat generated by the absorption of light. For this purpose, optical fiber is introduced into the lesion through a stereotactic procedure and local anesthesia. The technique is compatible with magnetic resonance imaging. It uses laser beam energy to heat and destroy tumors. TTIL works because certain nanoparticles can absorb the energy of a laser and convert it into heat. If the nanoparticles are hit by the beam while inside the tumor they release the energy with high temperature and destroy the tumor cells. The light can be focused into thin optical fibers; these can be inserted deep into the tumor mass, or into natural body cavities, in a minimally invasive way. The aim is to destroy the tumor tissue by thermal coagulation. (5)

Laser-induced thermotherapy is increasingly used for liver metastases which, due to their location or due to previous diseases, cannot be treated conventionally, i.e. by surgical removal. This procedure does not replace established procedures such as operations and/or chemotherapy, but should represent a more effective option in the concept of interdisciplinary therapy. In the case of already systemic diseases, the LITT local thermal ablation procedure can be performed in addition to the necessary chemotherapy and optimize the success of the treatment (6).

Advantages and disadvantages

The advantages and disadvantages of the LITT laser technique are shown in Table 1.

Table 1: Advantages and disadvantages of LITT lasers

Advantages	Inconveniences
More patient-friendly therapy	Norecommendableon extrahepatic metastases
Prognosis similar to other techniques	Number of injuries not exceeding five
Possibility of repeated use	Maximum lesion size is 5cm

THE INDICATIONS

If surgery is not possible, ILT is a minimally invasive, palliative and potentially curative option. In selected patients with HCC, liver metastases from colorectal cancer and other types of liver malignancy, it can improve survival to a similar degree as surgical resection. ILT makes it possible to treat a greater proportion of patients with liver malignancy than surgery alone, but a thorough understanding of the principles of treatment is essential to maximize the use of this technology. The extent of tissue necrosis is variable and depends on the laser device, type of fiber used, power settings, exposure times and tumor biology. Necrosis may be incomplete, particularly in tumors exceeding 5 cm in diameter. Continued refinements in laser technology, a better understanding of the mechanisms involved in laser-induced tissue injury, and innovative methods to manipulate these mechanisms may allow ILT to replace surgery as the primary treatment for liver malignancy in selected patients. Current indications include:

- Unresectable disease due to anatomic location of tumors or poor functional hepatic reserve and no evidence of extrahepatic disease.
- Patient preference
- Medical contraindications for surgery (7)
- no more than five lesions are present and no lesion exceeds 5 cm in maximum diameter (8).
- ILT can be used in combination with surgical resection to achieve complete tumor removal (9).

Complications

Complications of LTI were described by Mack et al. (7) with pleural effusion (7.3%) and intrahepatic abscess (0.4%) being the most common. Other lesions are described in the table (10, 11) Clinically relevant complications occurred in less than 2% of the patients.(Table 2).

Table 2: Complications of interstitial laser therapy LITT

Minor complications	Major complications
pleural effusion 7.3%.	pleural empyema 0.1%
intrahepatic abscess 0.4%.	intrahepatic bleeding 0.2%
subcapsular hematoma 3.1%.	intra-abdominal bleeding 0.2%
local infection in puncture site 0.2%.	bile duct lesion 0.1%
	tumor seeding

Intrahepatic abscesses, effusions and pleural empyemas may require drainage. There may be tumor seeding of the tumor, up to 20% develop it along the biopsy trajectory; 2 cases of tumor seeding after laser LITT have been described, which makes this an extremely rare complication of this procedure (12).

It has been shown that laser-induced interstitial thermotherapy can help achieve survival rates similar to those seen with surgical resection in liver metastases from colorectal and breast cancer, as well as other abdominal tumors (13), although randomized studies are scarce (14). The largest published series of any percutaneous ablative technique is laser ablation, the series consisted of liver metastases mainly from colorectal carcinoma (15).

BIBLIOGRAPHY .

1. Bown SG et al: Phototherapy in tumors. World J Surg 1983; 7:700-709.

2. Stafford RJ, Fuentes D, Elliott AA, Weinberg JS, and Ahrar K: Laser-induced thermal therapy for tumor ablation. Crit Rev Biomed Eng. 2010; 38:79- 100.

3. Rahmathulla G, Recinos PF, Kamian K, Mohammadi AM, Ahluwalia MS, & Barnett GH: Magnetic resonance guided laser interstitial thermal therapy in neuro-oncology: a review of its current clinical applications. Oncology. 2014; 87:67-82.

4. Vogl TJ, Straub R, Eichler K et al. Malignant liver tumours treated with MR imaging-guided laser-induced thermotherapy: Experience with complications in 899 patients (2,520 lesions). Radiology. 2002; 225:367- 377.

5. . Gilliams AR, Brokes J, Hare C. Follow-up of patients with metastatic liver lesions treated with interstitial laser therapy. Br J Cancer 1997; 76:31.

6. Mack MG, Straub R, Eichler K, et al. Percutaneous MR imagingguided laser-induced thermotherapy metastases. Abdom Imaging 2001; 26:369- 74.

7. Mack MG, Straub R, Eichler k, Engelmann k, Roggan A et al. *Percutaneous MRI image-guided laser-induced thermotherapy of liver metastases. Images of the abdomen* 2001; 26: 369-374.

8. Amin Z, Donald JJ, Kant R, Steger C.A, Bown SG et al. *Hepatic metastases: interstitial laser photocoagulation with real-time ultrasound monitoring and dynamic evaluation of CT treatment. Radiology* 1993; 187: 339-347.

9. Curley SA, cusack JC jr, Tanabe KK, Ellis LM. *Advances in the treatment of liver tumors. Curr Probl Surgery* 2002; 39: 449 - 571.

10. Vogl T.J., mack mg, Straub R, Roggan A, Felix R. Abdominal interventional radiology guided by magnetic resonance: laser-induced thermotherapy of liver metastases. Endoscopy 1997; 29: 577-583.

11. Vogl T.J., Eichler k, Straub R, Engelmann k, Woitaschek D et al. Laser-induced thermotherapy of malignant liver tumors: general principles, equipment(s), procedure(s)-side effects, complications, and outcomes. Eur J Ultrasound 2001; 13: 117-127.

12. Jorge A, Tarantino L, de Stefano G, Farella norte, Catalano O, Cusati B, et al. Interstitial laser photocoagulation under ultrasound guidance of liver tumors: results in 104 treated patients. Eur J Ultrasound 2000; 11: 181-188.

13. Vogl TJ, Naguib NN, Eichler K, et al. Volumetric evaluation of liver metastases

after thermal ablation: long-term results following MR-guided laser-induced thermotherapy. *Radiology* 2008; 249: 865-871.

14. Vogl TJ, Straub R, Eichler K, et al. Malignant liver tumors treated with MR imaging-guided laser-induced thermotherapy: experience with complications in 899 patients (2,520 lesions). *Radiology* 2002; 225: 367- 377.

15. Gillams Arkansas, Poso WR. Survival after percutaneous image-guided thermal ablation of colorectal cancer liver metastases. Dis Colon Recto 2000; 43: 656-661.

Buy your books fast and straightforward online - at one of world's fastest growing online book stores! Environmentally sound due to Print-on-Demand technologies.

Buy your books online at
www.morebooks.shop

¡Compre sus libros rápido y directo en internet, en una de las librerías en línea con mayor crecimiento en el mundo! Producción que protege el medio ambiente a través de las tecnologías de impresión bajo demanda.

Compre sus libros online en
www.morebooks.shop

info@omniscriptum.com
www.omniscriptum.com